John Herbert Aloysius Bone

Petroleum

And Petroleum Wells

John Herbert Aloysius Bone

Petroleum
And Petroleum Wells

ISBN/EAN: 9783744758895

Printed in Europe, USA, Canada, Australia, Japan

Cover: Foto ©berggeist007 / pixelio.de

More available books at **www.hansebooks.com**

PETROLEUM,

AND

PETROLEUM WELLS.

WHAT PETROLEUM IS, WHERE IT IS FOUND, AND WHAT IT
IS USED FOR; WHERE TO SINK PETROLEUM
WELLS, AND HOW TO SINK THEM.

WITH

A COMPLETE GUIDE BOOK

AND

DESCRIPTION OF THE OIL REGIONS

OF

PENNSYLVANIA, WEST VIRGINIA AND OHIO.

By J. H. A. BONE.

New York:
AMERICAN NEWS COMPANY.
PHILADELPHIA: J. B. LIPPINCOTT & CO.
1865.

PETROLEUM,

AND

PETROLEUM WELLS.

PETROLEUM, ITS DESCRIPTION AND HISTORY.—HOW IT IS FORMED AND WHERE IT IS FOUND.

What Petroleum is, where it is to be found, and what are the causes of its formation, are subjects now engaging the attention of the civilized world, and to neither of these questions have perfectly satisfactory answers yet been given. The name itself is from the Latin *petra*, a rock, and *oleum*, oil, being in fact "rock oil," deriving its name from being found in the rocks, or oozing from them. In its natural state its composition is very indefinite, consisting of various oily hydro-carbons, holding in solution paraffine and solid bitumen, or asphaltum. In some scientific works the fluid petroleum is described under the name of "naphtha oil," whilst that having a large proportion of asphaltum, is known as "bitumen." The latter is of comparatively little value, but the fluid petroleum, since the discovery of its manifold and important uses, has risen to be one of the most important staples. There appears to be no limit to its usefulness. The lighter oil, cleansed and purified, has come into almost universal request as an illuminator, surpassing all others, except gas, in brilliancy, and also possessing the merit of cheapness. The secret of producing gas itself, equal in illuminating power to the best coal gas, produced with much greater ease and at less expense, has been discovered and put into practice ; whilst, to show the capabilities of petroleum as an illuminator, the solid residium of the refining process is made into paraffine candles. As a lubricator for wheels and machinery the heavier qualities of petroleum have come into general use. Paint oils and varnish are made from it, and the benzine is used as a substitute for turpentine. Petrolized soap is a favorite toilet article. The most beautiful and durable colors and shades now in wear are obtained from the waste petroleum after refining. It has been used with success as a substitute for fish oil in tanning. For generations it has proved a valuable medicine applied both externally and internally. In fact, there seems to be no limit to its usefulness, for new applications of it are frequently discovered.

Petroleum, in one form or another, has been known in all ages, and in nearly all parts of the world, although many of its uses are the discoveries of the past few years. It is mentioned by the ancient Greeks and Romans, being known to the latter under the name of "bitumen." At Zante, one of the Ionian Islands, is an oil spring, still flowing, which was mentioned by Herodotus, more than two thousand years ago. In Sicily the ancient inhabitants burned petroleum in their lamps insted of fish oil. In the north of Italy it has for nearly two centuries furnished material for lighting the streets of Genoa and Parma. On the shores of the Caspian Sea, at Bakoo are extraordinary manifestations of petroleum oil and gases. These extend over a tract of country about twenty five miles in length, and about half a mile wide, in strata of a porous, argillaceous sandstone, belonging to the tertiary period. In the vicinity are hills of volcanic rocks, through which springs of the heavier sorts of petroleum flow. Open wells, from sixteen to twenty feet deep, are dug, and in these the oil gathers as it oozes from the strata. A large amount is annually gathered and distributed over Persia, where it is exclusively used for illuminating purposes, and for the sacred fires. The Rangoon district, on the Irrawaddy, is also famous for its large product of rock oil, and for centuries the whole Burman empire has been supplied with oil from this source. The annual yield of petroleum from this district is said to be more than 400,000 hogsheads, or about two thirds of the export from New York for 1864. The number of wells is 520. The natives use the oil as a medicine, burn it in their lamps, and grease timber with it to prevent the destructive operations of insects. Some of the Burmese oil has been sent to England and used in the manufacture of paraffine candles. In consistency it resembles the heavy lubricating oils of Pennsylvania and Ohio, whilst its color, of a greenish brown, is more like that of the lighter Pennsylvania oil. Petroleum is frequently found in the neighborhood of volcanoes. Around the volcanic isles of Cape Verde it is seen floating on the water; and to the south of Vesuvius a spring of it rises through the sea.

But it is in America that the largest deposits of liquid petroleum are found. Besides the principal reservoirs in Northwestern Pennsylvania, there are other deposits, the full value of which have not yet been ascertained, in Southwestern Pennsylvania Ohio, Western Virginia, Kentucky, New York, Canada, Kansas, and even California.

The cause of formation of petroleum and its location in the rocks, are questions that have as yet received no satisfactory solution. According to some geologists the oil originates in the coal beds, from which it is expelled by pressure, whilst others assert that the coal is formed from the oil, instead of the oil from the coal. In support of both of these theories the general resemblance of petroleum to the oil obtained from the distillation of coal is adduced, although there are some minor points of difference. But the existence of petroleum does not depend on the existence of coal in the same locality; on the contrary, the most productive oil districts are removed from the coal fields. In the Pennsylvania oil region the wells are entirely outside of the coal field, and so remote from it that there can scarcely be any connection between the oil and coal beds. The strata in which the oil is found dip south, and pass below the coal measures from five hundred to one thousand feet, the nearest coal bed to the more northern oil wells capping the highest hills about thirty miles distant.

Other geologists attribute the production of the oil to the slow distillation of animal or vegetable matter overwhelmed by ancient floods, and imprisoned in the rocks formed from the sand or mud in which the organic remains were buried. This theory presupposes an immense deposit of animal or vegetable matter, as the yield of oil has already been very large, and but a small portion of the deposit has been developed as yet. Another theory accounts for its production by volcanic agencies, but it is not by any means confined to the volcanic rocks. Some are disposed to look on it as a formation of by-gone ages, by processes long since terminated, whilst others, with a belief in the doctrine that Nature never stops in her work, assert that the process of formation is still going on, and that the supply is inexhaustible. An apparent confirmation of this opinion is found in the fact that the wells of Bakoo and Rangoon are as productive now as they were centuries ago. Single wells have dried up, but new ones have been sunk,

and the product of the district suffers no diminution.
This fact should allay the fears of those who are ap-
prehensive that the American oil regions will soon
be exhausted.

Petroleum is found in different parts of the world
in all the stratified rocks, and in the volcanic and
metamorphic formations. It is sometimes traced to
beds of lignite, and sometimes its source cannot be
discovered. In the United States and Canada the
sandstones are the most productive of oil. In the
Pennsylvania oil region the hills are capped with
conglomerate, lying in geological succession next below the coal measures. Through this the well is
bored, passing through alternating layers of shale
and sandstone, and terminating in sandstone, where
cavities exist, frequently filled with oil, gas, and salt
water. The dip of the strata in N. W. Pennsylvania
is nearly south. In Ohio it is east of south. The
most productive oil bearing sandstone crops out in
Ohio a few miles west of Cleveland, and dips gently
towards the Alleghany river, descending more rapidly as it gets farther south. In some parts of Oil
Creek, and on the Alleghany there are appearances
of a slight upheaval, forming cracks and fissures in
the rocks, and it is here that many borers look most
hopefully for oil in large quantities. According to
Prof. Evans, of Marietta, who has given the matter
much study, the oil is contained in cavities or fissures
in the rocks, in connection with both water and gas.
These are arranged, of course, according to their
weight, the water at the bottom, the oil floating
thereon, and the gas (often strongly compressed) fills
the upper part of the cavity. If such a cavity runs
obliquely from above downward, a well, when bored,
may strike either the water or the oil, or it may enter the gas chambers. In the first two cases, if the
gas be compressed, as it usually is, there will be a

spouting well—the water or oil, or both together, being thrown out of the mouth of the boring. When the tension of the gas is exhausted, resort must be had to pumping, until the cavity is pumped out. But in some cases a series of cavities communicate by small openings or crevices, in which case a well may flow intermittently for a long time, as it is replenished by percolation through these channels. It is not uncommon for intermittent wells to throw out at first 300 or 400 barrels a day, or to yield, in all, 20,000 bbls. They sometimes run two or three years before exhaustion. When there is little or no gas, or where, from the gas chamber being tapped, the gas is lost, pumping has to be resorted to from the first. Oil wells commonly vary in depth from 100 to 800 feet. Oil coming to the surface in springs is not a reliable sign of oil cavaties in the immediate neighborhood, for it is often carried a long distance by the current of the subterranean streamlets by which the springs are fed.

The oil of different districts varies considerably in specific gravity, and consequently in value. The lighter oils are more valuable for the purpose of illumination, and the heavier for lubricators. The Oil Creek petroleum is usually about 46° by Baume's hydrometer, being the lightest oil found. At some of the wells it increases in density to 38° At Tidioute, on the upper Alleghany, the Economite well oil ranges about 43° At Franklin the range is from 33° to 36°, and on French Creek and Sugar Creek the oil is also heavy, and is valuable as a lubricator. The heaviest oil is found at Mecca, O., the density being 26° to 27°, and the oil so thick that it will not flow in very cold weather. It bears a high price from its value as a lubricator. The oil obtained at Liverpool, O., is of a similar character. The heavy oils are usually found in comparatively shallow wells,

ranging from seventy to one hundred and eighty feet,
whilst the lighter are commonly found several hundred feet below. The "third sand rock" of the Venango County system, in which the largest deposits of
light oil are found, lies at a depth ranging from three
hundred to twelve hundred feet. The majority of
productive wells that have reached the third sand
rock, range from four hundred to six hundred feet
deep.

The yield of wells producing heavy, or lubricating
oil, is generally much less than the average of successful wells of lighter oil, but, on the other hand, the
value of the oil is much greater. A five barrel well
of Mecca oil is equivalent in value to at least a twenty barrel well on Oil Creek. This fact must be borne
in mind when comparing the value of wells in different localities. The "flowing wells" of large capacity
run the lighter grades of oil, the heavy oils requiring
to be pumped.

With regard to the condition of surface beneath
which oil is most likely to be found in paying quantities, there is as much difference of opinion as there
is in relation to the formation of the oil. In some
places on Oil Creek, for instance, wells sunk on the
flat bottom land are the most productive, and those
sunk in the side hill, or near it, find but little oil,
whilst on the next tract the reverse of this becomes
the rule. In and near Cherry Run several wells
have been sunk far up the steep bluffs, and have
proved successful. There appears to be no rule in
the matter without a large number of exceptions.

THE HISTORY OF PETROLEUM ON OIL CREEK.

The existence of oil in the valley of Oil Creek, in Venango County, Pennsylvania, was known for very many years. The Indians, from time immemorial, resorted to the valley at stated seasons to gather the oil for medical purposes; and the work of procuring it was prefaced and concluded with dances and other ceremonies. The oil bubbled up in mid stream in many places, and was obtained by throwing a blanket on the water, and, after it became saturated, squeezing the oil into the vessels prepared to receive it. The early settlers also used it as a medicine in cases of rheumatism, and it was frequently sold in druggists' shops for the same purpose, under the name of "Seneca Oil." An article in the "Massachusetts Magazine" for July 1791, describes the oil springs in what was even then known as Oil Creek, and says that the American troops, in their marching that way, halted at the spring, collected the oil, and bathed their joints with it. This gave them great relief, and freed them immediately from the rheumatic complaints with which many of them were affected. The troops also drank freely of the waters, which operated as a gentle purge.

About twelve years ago some attention was directed in different parts of the world to the subject of petroleum, or rock oil, and search was made for it in various directions. Among other places Oil Creek became the object of attention, and a company was formed to procure oil from the oil spring, the existence of which had become known to a large number of persons. Nothing was done, however, until in 1858, Col. Drake, of New Haven, Connecticut, visited the valley, and set about sinking a well on Watson's Flats, about a mile and a half below Titusville. The first well was unsuccessful, and another was sunk This was a success. The drill struck an oil cavity at a depth of seventy-one feet, and, on the tools being withdrawn, the oil rose to within five inches of the surface. It was pumped off, and yielded at first four hundred, and afterwards a thousand gallons of oil per day.

As may be imagined, the excitement in the valley was very great. Every one that held land in the vicinity of the Drake well made preparations for sinking wells on his own account, or leased to others a right to sink wells, reserving to himself a royalty of from one-eighth to one quarter the oil. Derricks were hastily put up, and "spring poles" fixed, all of the early wells being sunk by hand. Some of the wells were successful, but by far the larger portion obtained no oil at all, or in such small quantities as to be unremunerative. The demand was small, the use to which the oil was put being as yet very limited. Still, several of the adventurers were making fair wages, when the discovery of flowing wells revolutionized matters. Pumping oil at the rate of five to twenty barrels a day was a discouraging process when, at another well, the oil was running spontaneously as many hundreds as the others were pumping single barrels. The glut of the market, caused by the flowing wells, and the consequent depression in prices, rendered the continuance in operation of the pumping wells a losing speculation, and nearly all of them were abandoned. The lessees fled in despair, in many instances leaving their machinery behind them, and not stopping to surrender their leases. Some of the abandoned wells have since been success-

fully worked, and more would be, but from the im-
possibility of getting at the holders of the old leases,
and the fear to commence operations lest, at an un-
seasonable moment, the lessees should return.

The first flowing well ever struck was on the Mc-
Elhinney or Funk Farm, and was known as the Funk
Well, Funk was a poor man when the well was
sunk. It was struck June, 1861, and commenced
flowing, to the astonishment of all the oil borers in
the neighborhood, at the rate of two hundred and
fifty barrels a day. Such a prodigal supply of grease
upset all calculations, but it was confidently pre-
dicted that the supply would soon cease. It was an
"Oil Creek humbug," and those who had no direct
interest in the well looked day after day to see the
stream stop. But, like the old woman who sat down
by the river side to let the water run itself out that
she might cross dryshod, they waited in vain. The
oil continued flowing with but little variation for fifteen months, and then stopped, but not before Funk
became a very rich man.

But, long before the Funk had given out, the wonder in regard to it was overshadowed by a new sen-
sation. Down on the Tarr Farm the Phillips Well
burst forth with a stream of two thousand barrels
daily. Not to be outdone by the territory down the
Creek, the McElhinney tract " saw " the Tarr Farm,
and "went it a thousand better." The Empire Well,
close to the Funk, suddenly burst forth with its three
thousand barrels daily, a figure subsequent flowing
wells, vainly endeavored to equal.

The owners were bewildered. It was truly "too
much of a good thing." The real value of petroleum
had not yet been discovered, and the market for it
was limited. Foreigners would have nothing to do
with the nasty, greasy, combustible thing. Our own
people were divided in opinion. Some thought it a

dangerous thing, to be handled at arm's length,
whilst others set it down as a humbug in some way
or other, of which the community should keep as shy
as possible. The supply was already in advance of
the demand, but the addition of three thousand barrels a day was monstrous and not to be endured.
The price fell to twenty cents a barrel, then to fifteen, then to ten. Coopers would sell barrels for
cash only, and refused to take their pay in oil or in
drafts against oil shipments. Finally it was impossible
to obtain barrels on any terms, for all the coopers in
the surrounding country could not make barrels as
fast as the empire could fill them. The owners were
in despair and tried to choke off their confounded
well, but it would not be choked off. Then they built
a dam around it and covered the soil with grease, but
the oil refused to be dammed, and rushed into the
stream, making Oil Creek literally worthy its name.
For nearly a year it flowed, and then dropped to a
pumping well, yielding about a hundred barrels.
Lately it stopped, but on the application of an air
pump, it revived, and is now steadily increasing its
product.

The Sherman Well, which was the next great
"flowing well," was put down in the spring of 1862.
It was sunk under great difficulties. J. W. Sherman,
who was the original owner, commenced sinking it
on the Foster Farm, next above the McElhinney,
with very limited means, his wife furnishing the
money. After a while it became necessary to procure
an engine, but there was no money to make the purchase, and two men, who were in possession of the de-
sired article, were admitted to a share for the engine.
Soon after, when but a few feet more were necessary
to reach the supposed deposit of oil, the funds were
exhausted. A sixteenth interest was offered for $100,
and refused. Ultimately it was sold for $60 and an

old shot gun. A horse became necessary during the work, and a share was disposed of for the animal. At last, when all the means that could be raised by borrowing or selling were about exhausted, oil was struck, and flowed at the rate of fifteen hundred barrels a day. The flow continued at this rate for several months, when it declined to seven hundred barrels. For twenty-three months the well continued flowing, and then it stopped. For the first year the proprietors made but little, if anything, owing to the low price of oil and the difficulty of getting it to market, but, during the second year, the market improved, and an immense fortune was realized. The well now pumps from thirty to forty barrels daily.

On the East side of the Creek from the Foster Farm is the Farrel Farm. Farrell was a poor man, employed in hauling oil, and was offered one-eighth interest in the land for $200. In March, 1863, the Caldwell well was struck on that farm, not far from the Sherman well, and flowed twelve hundred barrels daily. Two months afterwards, the well now known as the Noble and Delamater, but then as the Farrel well, close to the Caldwell, struck oil, and commenced flowing at the rate of two thousand barrels daily. The column of oil spouted up fifty feet, with a roar like that of a hurricane. For some days the oil ran to waste, there being no possibility of controlling its flow. As soon, however, as its first fury was spent, a stop-cock was put on, and the flow reduced to a stream of the dimensions of a two and a half inch tube.

In the early days of oil enterprise, and after the yield had become large, considerable difficulties existed in getting the oil to a market. There were no railways to carry it off, and the only plan was to float it down the Creek to the Alleghany, and ship it thence by steamer or flat boat to Pittsburgh. When the Atlantic and Great Western Railway was built to Meadville, a large number of barrels were hauled across the country by teams to that place, and shipped thence to New York.

The supply of flat boats on the creek and river was far too small for the requirements of the oil trade. When boats could not be had the oil barrels were formed into a raft and lashed together. In this way they were floated, or towed, down to the mouth of the creek, where they were either loaded on steamers or towed to Pittsburgh. At times the great need was barrels. When this was the case the flat boats were made oil tight, and the oil poured into them in bulk. When there was not sufficient water in the creek, a large dam was made, and at an appointed time a pond freshet swept boats and rafts down to the river. Sometimes amusing, but expensive, casualities resulted from these pond freshets. An unskillful boatman occasionally got his boat in the wrong position, and the whole mass of boats, rafts, and floating tanks were thrown into confusion. Rafts were broken up, tank boats stove in, and an immense amount of property destroyed.

A far more dreadful disaster more frequently happened. With the oil from flowing wells a large amount of highly inflammable gas escapes. The utmost precaution is generally taken to prevent any fire being brought into contact with the gas, but accidents are sometimes unavoidable. The first rush of gas, on a flowing well being struck, occasionally enters the engine shed, and takes fire from the furnace. A terrific explosion follows, and every thing in the vicinity is wrapped in flames. When a well takes fire in this way it is very difficult to extinguish it. Water appears to have no effect, the only effectual way to extinguish the flames being to turn on steam, or stop the well hole by throwing dirt on it—

a rather difficult task to perform in the near presence
of an intensely hot "pillar of fire." Several of the
leading wells have been on fire, and much damage
done. Several times the boats on the creek took fire,
and, breaking from their moorings, swept down
stream, carrying devastation with them. A terrible
fire of this kind occurred May 12th, 1863, when a
great number of boats, loaded with oil in bulk and
barrels, were on fire, and endangered the existence of
Oil City. They swept down the Alieghany, destroy-
ing every thing with which they came in contact.
The bridge across the Alleghany at Franklin was
totally consumed. The construction of railways to
the oil district, from Cory and Meadville, by lessen-
ing the necessity for boats and rafts, have greatly
diminished the risks by freshets and fires.

In briefly sketching the history of a few of the
flowing wells, only those of the earlier and more famous have been selected. A number of other flowing wells have made their possessors wealthy, and
some have attained considerable notoriety.

The change in the fortunes of the original owners
of property in the oil regions must be a source of
wonder to themselves, as it is to every one else. The
so-called "farms" on Oil Creek never produced
enough to give decent support to those who lived on
them. The residents on the creek led a rough life,
generally eking out their livelihood by rafting lumber to Pittsburgh, and bringing from that city such
articles as they needed. So poor were many of them
that they were compelled to foot it home from Pittsburgh, for want of means to pay for a conveyance.
The revenue derived from oil leases on their lands,
and fortunate speculations in oil and territory, have
made all of them wealthy, many of them millionaires.
Land that six years since was not worth ten dollars
an acre, has in some instances brought as many
thousands.

Until recently, the wells in the Pennsylvania oil regions were owned by single adventurers, or by a few men associated together. The disadvantages in this mode of working were many and great. An adventurer who found his money and labor expended in the production of a "dry hole," rarely possessed means, or perseverance enough, to sink another well in the neighborhood. The well was abandoned, and gave a bad reputation to the whole neighborhood. With the present systems of joint stock companies, able to prosecute their work in spite of two or three failures, old property is more thoroughly developed, and the merits of new oil bearing territory properly tested. The individual profits are not always so great, nor are the individual failures so ruinous. Oil mining becomes less a game of chance, and takes its place among those branches of business that offer a good prospect of profitable returns for the investment made in them.

The growth of the petroleum business is indicated in some degree by the following summary of exports in 1862-3-4. The exports in 1861 were small and no accurate account was kept of them. It must be borne in mind that the consumption in the United States is very large and rapidly increasing; in fact, petroleum has become almost a necessary of life with us.

TOTAL EXPORT IN 1864, 1863, AND 1862.

	1864. Gallons.	1863. Gallons.	1862. Gallons.
From New York.......	21,335,784	19,547,604	6,720,273
Boston	1,696,307	2,049,431	1,071,375
Philadelphia.....	7,760,148	5,395,738	2,800,973
Baltimore	929,971	915,866	174,830
Portland	70,762	342,062	120,250
Total export from the United States........	31,792,972	28,250,721	10,887,701

There was also exported, in 1864, from Cleveland direct to Liverpool 80,000 gallons refined.

AVERAGE PRICES FOR 1864 AND 1863.

	Crude.	Refined, free.	Refined in bond.	Naphtha refined.
Average 1864....	4,181	7,461	6,503	3,954
Agerage 1863....	2,813	5,174	4,415	2,853

HOW OIL WELLS ARE BORED AND WORKED.

The individual or company intending to bore for
oil, either purchases the land in fee simple, or obtains
an "oil lease." At the present time the purchase
of land in fee simple is mostly effected by companies,
the high price put on oil bearing lands rendering it
almost impossible for individuals to obtain a tract of
any considerable dimensions.

Of the tract thus purchased but a small proportion, generally, consists of what is now considered
"borable territory," namely, the flat land bordering
on the river or creek, and sides of a ravine, or bank
of a stream. The remainder is usually high bluffs,
valuable in proportion to the amount of wood obtain-
able for fuel. Some companies work their own property, whilst others grant oil leases to individuals or
other companies.

An "oil lease" grants to the lessee a right to bore
within certain limits for "oil salt or other minerals,"
the work to be commenced within a given time, and
"to be prosecuted with all reasonable dilligence." If
these conditions are not complied with the land re-
verts to the owner of the fee simple. Some of the
leases granted in 1860 and 1861 were loosely drawn
up, and made no provision for the reversion of the
property in case of the abandonment of the works,
and it is no unfrequent thing, after an old well has
been taken by new adventurers and made successful,
for the original lessee to make his appearance and
claim compensation. If the new proprietors do not
comply, an injunction is obtained, and, rather than

have the work stopped for months until the case comes up for trial, the victims are generally willing to compromise. In the beginning of the oil enterprises on Oil Creek, as now in some of the new oil territory, the owner of the fee obtained, as compensation for granting the lease, a royalty, or "landed interest," of one sixth or one-fourth the oil raised, leaving the remainder, or " working interest," to bear all the expenses. At the present time, on Oil Creek, the landed interest obtains one half the oil on all new leases, and in some very desirable locations a bonus is demanded in addition. It will be readily seen that the owner of the "landed interest" gets the lion's share, receiving half the products without being at any expense for working. The royalty was formerly paid in kind, but is now usually settled by taking half the value of the sale of oil. Should the lessee abandon his adventure he is allowed to remove his derrick and engine.

Having bought or leased a location, the next step is to select the exact point for boring. In this the experienced worker is guided, to some extent, by the nature of the soil and the position of the ground. A new class of people has sprung into existence under the cognomen of "oil smellers," who profess to be able to ascertain the proper spot for boring by smelling the earth. Some of them practice considerable mummery in order to mystify and impress their employers. The "witch hazel" is also frequently used, the professional locater of wells marching solemnly along, holding his hands apart with one end of a forked hazel in each. On passing over an oil spring or basin, the point at the junction of the forks suddenly deflects towards the earth, and there the work is commenced. As the witch hazel has the property—according to believers in its powers—of finding streams of water in the same manner, it sometimes happens that water, instead of oil, proves to be the product of the well.

The exact spot being determined, a huge derrick is erected immediately over it. This is a square frame of timbers, substantially bolted together, making an enclosure about forty feet high, and about ten feet at the base, tapering somewhat as it ascends. This is generally boarded up a portion of the distance to shelter the workmen. A grooved wheel or pulley hangs at the top, and a windlass and crank are at the base. A short distance from the derrick a small steam engine, either stationary or portable, is fixed, and covered with a rough board shanty; a pitman rod connects the crank of the engine with one end of a large wooden walking beam, placed midway between the engine and the derrick, the beam being pivoted on its centre about twelve feet from the ground. The walking beam is a rude imitation of that of a side-wheel steamer. A rope attached to its other end . passes over the pulley at the top of the derrick, and terminates immediately over the intended hole. A cast-iron pipe, from 4½ to 5 inches in diameter, is driven into the surface ground, length following length until the rock is reached. In the older wells the ground was dug out to the rock, and a wooden tube put in it. The earth having been removed from the interior of the pipe the actual process of boring or drilling is commenced. Two huge links of iron, called "jars" are attached to the end of the rope. At the end of the lower link a long and heavy iron pipe is fixed, and in the end of this is screwed the drill, about three inches in diameter, and a yard long. When all is ready the drill and its heavy attachments are lowered into the tube and the engine set in motion. With every elevation of the derrick end of the walking beam, the drill strikes the rock, the heavy links of the "jars" sliding into each other and thus preventing a jerking strain on the rope.

The rock as it is pounded mixes in a pulverized condition with the water constantly dropping into the hole, and assumes a pasty form. After a while the drill is hoisted out and a sand pump dropped into the hole. The sand pump is a copper tube, about five feet long, and a little smaller than the drill, having a valve in its bottom opening upwards and inwards. As the tube is dropped into the hole the pasty mass rushes into it through the valve and remains there. When this has been done several times the tube is hoisted out and emptied, the operation being repeated until the hole is clear, when the work of drilling recommences. It is evident that as the drill is not round at the point, but with a chisel shaped edge, the hole would not be round unless some other means were adopted. This is partially accomplished by the borer, who sits on a seat about six or eight feet above the hole, and holds a handle fixed to the rope, giving the latter a half twist at every blow. By this means a nearer approach to a cylindrical hole is attained. But the hole must be as nearly round as possible, and therefore the tools are taken out, and a "rimmer," or "reamer," sent down which cuts down the irregularities of the hole.

In the earlier days of well boring, as now in some localities, the wells were sunk by hand, or by horse power. In the former case a stiff spring pole, firmly secured at one end, lifted the drill and rods suspended from its free end, and the power was applied to this end to make it suddenly descend. Two men, standing together, placed each a foot in a double stirrup suspended from the pole, and suddenly bore it down. Immediately it sprang up, and the operation was repeated. This was a tedious and laborious operation, and has been generally abandoned.

As the holes get down to points where the first indications of oil are reached the contents of the sand pumps are anxiously examined. The oil borers have a geological system of their own, the prominent points of which are three layers of sandstone. The "first sandstone" lies immediately blow the alluvial deposit. The "second sandstone" is at a variable depth of 100 to 300 feet, and here the first indications of oil was reached. Some wells go no lower than the second sandstone, but the general plan is to go down into the "third sandstone," where the largest and most reliable deposit of oil is found.

It frequently happens that the drill breaks and falls off, and becomes fixed in the hole. Nothing can be done until the tool is removed. The remaining portion of the boring instrument is taken off, and a pair of nippers or clamps let down into the hole to grip the broken drill and extract it. Some men make the extraction of tools a special business, and exhibit great ingenuity in their devices to overcome the difficulties they have to encounter. There are instances where wells have had to be abandoned in consequence of the tools remaining immovably fixed in the hole.

When the hole has been sunk to a sufficient depth and "strike ile," the next thing is to extract it from the well. If a flowing well has been struck all trouble on this head is saved, as the oil and gas rush out in a stream, sometimes with such violence that the men have to make their arrangements with considerable rapidity, or the precious fluid runs to waste. If, on the contrary, it is a pumping well, an iron pipe, with a valve at the bottom like the lower valve of a pump, is run down the entire depth of the well, the necessary length being obtained by screwing the sections firmly together. A pump box, attached to a wooden rod, also made of sections screwed into each other, is inserted in the tube, and the upper end of the rod attached to the "walking beam." The well is now ready for pumping.

One important feature in the tubing process must not be forgotten. In boring for oil, springs of water are of coarse cut through and the water falls into the hole. Being heavier than the oil it lies at the bottom, and would enter the pump tube but for a very ingenious contrivance known as the seed-bag. This is a leather bag, in shape something like a boot leg, filled with flax seed, which is fastened around the iron tube at what is considered the proper point, and crowded down with it. When the seed-bag becomes wet it swells and thus forms a water tight packing between the tube and the rock. At times the seed-bag slips or bursts, the well at once fills with water, and the tubing has to be pulled in order to refix the seed-bag.

More or less gas accompanies the oil in its passage to the surface. If a flowing well, the gas is allowed to escape, there being no use for it, and it can be distinctly seen puffing out of the pipe, generally with labored breathings or panting, the cause of which is known among the operators as the "breathings of the earth," in reality being the irregular obstructions to its passage by the unequal flow of oil in the bottom of the hole. The passage of the oil from a large flowing well is a curious and interesting sight.

In many of the pumping wells the gas is saved and used, either by itself or with coal, as fuel for the engine. To save it, the mingled gas, oil and water, —for in spite of all precautions some water will come up from nearly every pumping well—is conducted by a pipe from the well tube into a tight barrel. The oil and water fall into the bottom of the barrel, and run off by a pipe near the bottom into a huge tank or vat, where another separation is caused by the different gravities of the two fluids, the water sinking to the bottom of the vat. The gas escapes by a small pipe at the top of the barrel, and is conducted into the furnace, where it burns with a fierce and steady flame. The engine of the Forest City well, as also many other wells, is run entirely by gas, the jet being spread into a broad and waving flame by passing through a piece of sheet iron pierced with holes. Its steadiness is shown by the fact that the engine house is lit with several jets of gas, of a steadier and purer flame than that furnished by some gas companies.

The oil, as it flows into the tank, is a dark green fluid. When sold for shipment it is drawn off by a faucet in the bottom into barrels. In the larger wells, where a considerable quantity of oil is kept on hand before sale, ranges of vats are built, the oil flowing from one to the other. The vats are covered with boards, and at the larger wells roofed in to prevent evaporation. At the gassy wells great care has to be taken with regard to fire, as a lighted cigar might set fire to the gas and blow up the whole concern. In the early days of the flowing wells, before their nature was thoroughly known, serious conflagrations took place from this cause. Should a well take fire, water not only fails to extinguish it, but seems to add to the fury of the flame.

A new process of boring is on trial at the Gillette Company's wells, on the McElhiney tract, under the management of Mr. J. T. Briggs. The process is of French invention, and the patentee personally superintends its working. This is the first time it has ever been tested, and the progress of the experiment is watched with great interest by well owners. The principle is that of cutting out a hole instead of pounding it. The drill is circular and hollow, being a thin tube, set at its lower edge with Brazilian diamonds, of hardness sufficient to cut glass. It is connected by an iron rod to beveled cog wheels attached by cranks and rods to the walking beam of the engine. The surface of the upper rock being cleared, the drill sits on it and revolves with great rapidity. cutting its way down at a rate astonishing to old well borers, and leaving a central core standing. A clamp is let down which grips the core and jerks it up in the form of a perfectly smooth cylinder. Water is poured down the hole to assist the cutting process, until the natural flow from the springs cut supplies the want. The portions of the core shown exhibited the stratification of the rock, and will go far to settle some vexed questions about the strata which cannot be ascertained by the ordinary method of drilling.

Five feet of rock had been cut at the rate of four inches in five minutes, or ninety-six feet per day, when some changes were required in the machine, and it was removed for alteration. The patentee is satisfied that he can put down a well five hundred feet in ten days, at no greater cost to the well owner than by the present tedious process, which takes from two to four months.

It sometimes happens that after a well has been yielding for months its stops and refuses to yield another drop. This is occasioned in some instances by the thickening of the paraffine at the bottom of the hole, and the consequent obstruction to the flow of the oil into the pump box. To remedy this a jet of steam from the engine is introduced and forced down the hole, melting the coagulated mass, and restoring the flow into the pump. Another plan, which is coming into use, and which has so far proved successful at the Empire, is to use an air pump, with two pipes inserted into the tube of the well. The air is forced down one pipe into the vein at the bottom, and the oil rushes up in a steady stream through the ther. Both these processes are as yet but experiments, although the air pump has been so far suc-

cessful, and has proved much superior to the ordinary pumping process in producing an abundant and
steady yield.

The cost of sinking a well, including purchase of
engine, tools, and every necessary, ranges from six
to nine thousand dollars, according to location, depth
of well, size of engine, &c. The second well can be
sunk at less than half the expense, by using the
same engine and tools.

THE PENNSYLVANIA OIL REGION—THE PRINCI-
PAL OIL LOCALITIES AND HOW TO
REACH THEM.

Pennsylvania is the greatest oil producing State in America, or the world, and Venango county is the principal oil region of Pennsylvania. Some developments of oil have been made in Crawford, Clarion, and Fayette counties, but so far, Venango county has been the chosen seat of empire of King Petroleum. If a line be drawn nearly through the centre of the county, running from north to south, tending a little west, it will pass along Oil Creek, the central and most productive portion of the oil territory. From Franklin, a few miles below where Oil Creek joins the river, the Alleghany to the east, and French Creek to the west, form a huge V, with Oil Creek passing down the middle and joining the right arm of the V just above the point of junction. Below this point the Alleghany stretches, converting the V into a Y. The centre, or Oil Creek line, is that of the greatest yield at present, the others not having been so extensively worked. The first discoveries of oil were made on Oil Creek, and for some time explorations were confined to that line. The success there met with induced others to examine into the oil-producing qualities of the adjoining streams, and a number of holes were sunk on the banks of the Alleghany River, French Creek, Sugar Creek, (an affluent of the French,) and some of the "runs," or small streams, tributary to the several creeks.

Beginning at the point where the Alleghany crosses the Venango county line, a short distance below Tideoute, (the highest point of oil operations on the river) the oil line stretches along the banks of the Alleghany, in its sinuous course, for about sixty miles, during no part of which distance would a voyager be out of sight of the derrick of an oil well, past, present, or prospective. After entering the Venango county lines, the principal streams discharging into the Alleghany on its way south, are the East Hickory and Little Hickory on the east; West Hickory on the west; Tionesta and Little Tionesta, Hemlock Creek with its branch known as Porcupine Run, on the east; Culbertson's Run and Pithole Creek on the west; Horse Creek on the east; Oil Creek, Two Mile Run, and French Creek on the west; East Sandy on the east; Big Sandy, Big Scrub Grass and Little Scrub Grass on the west. Besides these there are numerous smaller streams that have not yet attained notice for their oil bearing qualities. All the streams mentioned have become oil locations, and on each of them the work of pumping or boring is going on with great activity.

From the mouth of Oil Creek to Titusville, just across the Crawford county line, is a distance of twenty miles, along the whole of which the wells are thickly planted. Ascending the stream, Cornplanter Run heads off towards the north west, Cherry Run to the north east, Cherry Tree Run, and Weikel Run which branches from it, to the north west; and Bennehoff's Run to the west. Just below Titusville, Oil Creek forks to the east and west. All the tributaries of Oil Creek are oil producing, and are crowded with wells.

French Creek, one of the largest affluents of the Alleghany in Venango county, comes in from the north west. A number of wells are scattered along

its banks. Sugar Creek enters French Creek from
the east about three miles above Franklin, and is
now a favorite oil locality.

Meadville is the central point of departure from
which to reach any part of the Venango county oil
regions. From it the traveler can enter Oil Creek
at either the Oil City or Titusville end. The best
route is by way of Franklin and Oil City. Arriving
at Meadville by the Atlantic and Great Western
Railroad, the visitor can obtain a comfortable night's
rest and an excellent breakfast at the McHenry
House, the hotel in the depot building. Taking the
Franklin Branch cars a little before eight o'clock in
the morning, the distance to Franklin, twenty-eight
miles, is done in something under two hours and a
half. The railway follows the course of French
Creek throughout, affording in summer a series of
picturesque scenes. The last five or six miles of the
route is lined with oil wells, nearly all put down in
1860 and 1861, and abandoned. A few have re-
sumed work.

From Franklin to Cooperstown, on Sugar Creek,
is about eight miles over a fair road. Conveyances
can be had in Franklin. A nearer route to Cooperstown is to leave the train at Utica station, nineteen
miles from Meadville, and take the road across.
This will save from two to three miles, but the
chances of obtaining conveyance across are not many,
as there are no livery stables in Utica.

Horses or conveyances of some kind can be obtained in Franklin to visit the Alleghany below, or
to reach Oil City. The visitor must not rely too
much on this, however, as the great rush of people
to the oil regions makes a greater demand for conveyances than can be met by the limited supply.
As might be supposed under such circumstances,
prices rule extravagantly high. Livery stable keep-

ers charge about ten per cent. on the value of a horse when letting it out for a day. The Alleghany below Franklin is very crooked, and the distance by the river bank is much greater than by the roads that keep at a short distance from the stream. Unless prevented by ice, the communication between Franklin and Oil City is kept up by a daily steam packet, the " Petrolia No. 2," which leaves Franklin on the arrival of the morning and evening trains, leaving Oil City on return, about 8 A. M. and 3 P. M. The Franklin Branch Railroad will shortly be open to Oil City, which is seven miles from Franklin.

From Oil City, up the Alleghany, there is a road hugging the river bank, and crossing the river by ferries at several points where the steep bluffs block the way. To reach Pithole Creek, and the river above that point, the best route is to go from Oil City to Plumer, seven miles, and turning to the right from the centre of the village, cross Pithole Creek, and strike the river at Culbertson's Run. From this point a good road extends along either bank. Two villages are passed on the road before reaching the northern line of Venango county. At President, near the mouth of Hemlock Creek, a large hotel has recently been built. At the mouth of Tionesta Creek is the village of Tionesta. Both these villages are on the east side of the river. From Oil City to Hickory Creek is about twenty eight miles.

From Oil City up Oil Creek to Titusville the choice lies between horse back and foot travel. The best way, on every account, is to walk. The horses betray few traces of Arabian blood, and their habits are too devotional for comfort or safety. The greater number drop on their knees at every opportunity. By going on foot the visitor can see more, and, in many instances, travel faster than on horseback.

From Oil City to Shaeffer's Farm, where the Oil
Creek Railroad is first reached, is about twelve and
a half miles of about the worst road—or rather no
road—in the United States. There are several stopping places on the route, Rouseville, McClintockville and Petroleum Centre having tavern accommodations—such as they are. At Shaeffer's Farm
the train can be taken in the evening for Titusville,
seven miles, and, early next morning, from Titusville to Corry, twenty eight miles, and back to
Meadville, forty two miles farther.

Every where the visitor must be prepared for
rough living and hard lodging, if fortunate enough
to obtain lodging. When intending to stop at night
at any particular place, telegraph in the morning to
engage a bed. By doing so you will have a slight
chance of obtaining half a bed. If this is neglected
there is a certainty of getting no bed at all.

Wear such clothing as will excite no regrets should
they be covered with mud or grease, as they inevi-
tably will. Put on long legged boots, made water-
proof. Carry no baggage except a small traveling
bag or haversack, suspended by a strap over the
shoulder. A blanket will be found very convenient
in case no bed can be obtained, or as an addition to
the scanty amount of bed clothing, should a bed be
secured. A lunch, or some crackers and cheese, in
the haversack will be found convenient in case a
tavern cannot be reached by dinner time.

The visitor to the Oil Regions who cannot "rough
it," amid mud, filth, grease, wretched roads, deep
quagmires, miserable accommodations and poor food,
had better stop at Meadville, eat a hearty dinner at
the McHenry House, and then take the first train
for home. He has not had a " call " for life among
the oil wells.

A TRIP DOWN OIL CREEK—MEADVILLE TO SHAFFER'S FARM, VIA CORRY.

In November and December, 1864, the writer spent three weeks exploring the oil region of Venango county, and investigating the condition of some of the oil properties. All the known oil localities were visited and carefully examined. A narrative of the leading features of the trip will give the reader some idea of the nature of the country and the business done in it, but no description can do the subject proper justice. An actual visit can alone give one a proper appreciation of the vast importance of the petroleum business.

Traveling in the roomy and elegant cars of the Atlantic and Great Western Railway, the journey was performed with comparative ease and comfort. At Meadville we halted for the night at the McHenry House, that we might enter the oily land with daylight to reveal its wonders. Here we found the principal topic of discussion was oil. The wave of excitement which was said to be sweeping through the valleys to the southward, rippled gently in the McHenry House, and people were discussing the latest news from "the Creek." Every one we met with was "in oil," and every one was making arrangements to get deeper into the grease. Big stories were told of the fortunes made at the wells, and by the owners of oil lands, and bigger tales of the frightful state of the roads. I dreamed all night of thousand barrel wells throwing up oceans of mud,

2

and wading in greenbacks to the knees. My traveling companion in the opposite bed interrupted his
"distinct breathings," with muttered offers of "ten
thousand dollars for the refusal of your farm for five
minutes," awakening me with his demands for an
immediate answer. I set him down as a pitiable
case of "oil on the brain," and tried to go to sleep.

At five o'clock of a dark, cold, and snowy morning, we set out by a freight train for Corry, having
determined to enter the oil region by the Titusville
route. That it is not the most convenient route was
a fact of which we soon had abundant evidence, but,
on the whole, there was not much to complain of, although traveling in a caboose car very early on a
cold morning, is not the most pleasant experience in
the world.

Everything must have an end, and shortly after
eight o'clock we reached Corry. Here the Atlantic
& Great Western, Philadelphia & Erie, and Oil
Creek Railways meet. The junction station is a miserable little affair, of rough boards, and utterly un-
able to shelter one half the crowd waiting to go by
the different trains. The snow was driving furiously, the weather was getting momentarily colder, and
every one sought shelter. The dense mass, packed
into the miserable little station like herrings in a
cask, formed a motley assemblage. There were but
few women among them. The men were of all
ranks, ages, and descriptions. Sharp eyed, trim
dressed, and eager speculators from New York,
Philadelphia and Pittsburgh, carpet-sack in hand, or
with traveling bag strapped over the shoulder, going
down to secure "a big thing;" traders anxious to
open up a line of custom; rough fellows, going down
to work at the wells; and old farmers, coarsely clad,
and with their cowhide boots covered to the tops
with mud whose layers spoke of months of travel

over villianous roads, just as the concentric rings of
bark on a tree reveal its years of growth, but who
had within a year or two been made rich by farms
that had previously made them poor:—all were
bawling for tickets for "Titusville" or "Shaeffer's
Farm," until the ticket clerk was well nigh driven
desperate.

For nearly an hour the crowd surged outward
towards the platform, as the rumble of a passing locomotive was heard, and inward towards the stove,
as the origin of the sound became known. Just as
the crowd had settled down to the conviction that
there was to be no conveyance to Oil Creek, a shout
of " Train" was heard.

A bomb shell suddenly dropped in their midst
could not have produced a greater stampede than
did that shout. The train was slowly backed down
to the station, when the crowd rushed furiously at it.
They swarmed up the steps, into the baggage car,
over the locomotive, everywhere but under the
wheels, and how they escaped that was a mystery.
All the courtesies and amenities of life were disregarded. Men fought for precedence as if their lives
depended on it. Women were rudely thrust back
by anxious men who clung to the step rails and
kicked off those who endeavored to climb over them.
Three cars and a baggage car were in three minutes
packed almost to suffocation. A rattle and a jerk,
and the train was off, shaking and jolting every one
into position. We were well on our way for the
" Oil Dorado."

From Corry to Titusville the railroad passes
through an irregular country, and the track generally follows the original configuration of the land.
Up hill the huge locomotive pants with its heavy
load, and down hill it rushes, shrieking as if anxious
to plunge itself into destruction. Corry, with its

scattered houses, its immense brick oil refinery, the largest establishment of the kind in existence, and all the other items that make up a thriving town, where three or four years ago there was nothing but "the forest primeval," is soon left behind. So also are the long trains of engines waiting to be united to innumerable derricks already lining the creek, but which will have to wait longer yet owing to the inadequate facilities possessed by the road for transporting the immense amount of freight crowding on it. Soon the line of the creek is struck, and the road skirts its edge, most of the way winding along a ledge cut in the face of almost perpendicular cliffs. Here and there a derrick, like the skeleton of a church spire with its apex sawn off, and the frame not yet lifted on the church, keeps solemn watch along the banks, pickets of the advancing army of Petrolia. Presently the derricks increase; they close up their ranks, and soon stand in unbroken line along the left bank of the stream, throwing frequent skirmishers across to the right bank, effecting lodgments at the foot of the precipitous cliffs, where there is barely room to stand, and even threatening the railroad track which winds higher up. Puffs of steam and creaking of engines show where the pumping wells are at work. The river is dark, and a scum of oil glistens on its surface. Here and there a small board shanty, connected by slender pipes with tanks at a little distance, marks the existence of a refinery—for all the processes connected with oil, from its extraction from the rock until it is ready for consumption, are carried on in the vicinity of the wells, employing a great number of refineries in addition to those in successful operation at Cleveland, Pittsburgh, New York, Philadelphia, and other places. The river margin widens and the number of derricks increase. No longer in single line, they double and treble their ranks, and appear in unbroken column; the new timber showing the large proportion just started, and the black and greasy appearance of many proving that their owners have "struck ile."

Twenty eight miles from Corry the train stopped at Titusville, the last point in Crawford county before entering Venango county. A few years ago, Titusville was a lively little village, chiefly inhabited by lumbermen and raftsmen. In 1855 it was credited in the Gazetteer with having "an universalist church and 243 inhabitants." Now it has a population of over six thousand, and rapidly increasing. The one church has found several others to keep it company. There are thirteen hotels, crowded nightly with guests, of whom a large proportion have to spend the night without the privilege of half a bed, (an entire interest in a bed is a thing unknown in the oil regions.) Two banks do a large business in the funds produced by operations in oil, and a third bank is nearly ready to open. A new and handsome reading room, well supplied with the papers and periodicals of the day, has been opened, and there is a hall kept constantly engaged by lecturers, concerts, or other popular amusements. In every part Titusville gives evidence of its state of transition from a small village to a thriving city. Lofty and handsome brick blocks alternate with small dilapidated wooden buildings. A well made plank sidewalk borders a muddy canal, by courtesy called a street. When the citizens have time, some day, they will probably rectify those little irregularities, but just now every one is too busy. There are oil wells within the limits of the town, and some of the new settlers who have purchased lots on some of the streets are in doubt whether to erect a dwelling or a derrick. One is necessary, yet the other may pay best.

The platform at Titusville station was crowded with people, some waiting to see the new arrivals, but most intending to take the train for farther down the creek. For every person who left the train at least three got on, so that the crowd became even thicker than before, and a number were driven to the platform of the cars.

From Titusville to Shaeffer's Farm is seven miles, and all the way there were abundant evidences of oil adventures, past, present, and prospective. About a mile and a half below Titusville, on Watson's flats, is the scene of Col. Drake's first experiment in sinking oil wells, the result of which has been the enriching thousands of persons, and the addition of an immense business to the resources of the nation. Near this point comes in the East Branch of Oil Creek, which has now been purchased and leased to nearly its entire length, for the purpose of boring for oil. Along the whole route to Shaeffer's Farm the derricks increased in number until there was a perfect forest of dismantled steeples. The air was redolent with the greasy perfume, and the passengers in the crowded cars talked more fiercely about oil, and discussed vast sums of money more glibly.

Miller's Farm, at which the train stops for a few minutes, is now a scene of busy activity. In the Autumn of 1864 three fourths of the farm, comprising a tract of three hundred and seventy five acres, was purchased by the "Indian Rock Oil Co." of New York. The enterprise was a vast one, for the purchase of so large a property in the very heart of the developed oil region, required large capital. Among the heaviest proprietors in the Company are William H. Webb, of New York, the world renowned ship builder, Ocherhausen Brothers, extensive sugar refiners of New York, Raynolds, Pratt & Co , wholesale druggists, in that city, and J. T. Briggs of Titusville, one of the pioneers of oildom. The property at the time of its purchase contained a number of wells in operation, and others going down. The new proprietors, anxious to fully develope the value of their property, instead of floating their stock upon the market, proceeded at once to the work of sinking new wells, and in this way have already expended about $75,000. The result of this course will, from present appearances, be the production of several new successful wells. Besides their property on the Miller Farm, the Indian Rock Oil Co. have a large interest in twenty five acres on the Foster Farm, lying farther down the stream. Altogether the Company have thirty wells—a very desirable property to hold. The President of the Company is G. A. Hoyt, manager of the Pennsylvania Coal Co.; and the other officers are, Victor L. Conrad, Secretary and Treasurer; J. T. Briggs, General Superintendent at Titusville; L. H. Severance, Assistant Superintendent.

Shaeffer's Farm at last. The crowd tumbled out of the cars as frantically as they clambered in, and, clutching their scanty baggage, rushed wildly for the "hotel," scrambling over each other in their anxiety to get first at the register. Every man as he scrawled his name with a nervous hand, enquired if he could get a room at night, and was met with the chilling response from an individual in high boots, covered with mud, that "there was almighty little show for anything, as it looked to him." Determined to make sure of what was at hand, and trust to luck for the future, the crowd broke for the dining room, not stopping to go through the ceremony of washing, for the land of grease and dirt has been reached, and the niceties of civilized life are henceforth disregarded. A plunge was made through the narrow passage to the dining room; already keen scented nostrils snuffed the titillating odor of roast and boiled, and hungry mouths watered in expectancy.   But an impassable obstacle presented itself in the shape of a grim janitor who refused admittance without a ticket, and the " Johnny Newcomes" had to fight their way back to the bar and deposit seventy-five cents for the bit of blue pasteboard, whilst the old stagers who were better provided, entered and filled all the vacant chairs.   To give an idea of the rush of pilgrims to the Oily Land the fact may be cited that at one tavern at Shaeffer's Farm, from two to four hundred people dine daily. As to the quality of the meal we have nothing particular to say.   The price was first class, and if the viands fell short of first class standard, the people at the "oil diggins" have no business with nice stomachs.

Dinner bolted, our first enquiry was about getting down the Creek.   Conveyances there were none, from the fact that there were no roads to travel on. A single glance at the country around the "hotel" settled that question.   A walk of three feet from the door in any direction brought the wayfarer into mud knee deep,—and such mud ! Clayey, slippery, greasy, sticky mud, into which you slid easily to uncertain depth, but which clung with fond affection to your legs, and endeavored to perform the offices of a boot jack ; deceptive mud, that appeared of uniform quality, but which in places suddenly engulfed the traveler thigh deep.   Some of the pilgrims struck out boldly but were soon stuck fast, monuments of their own rashness.   Clearly that mode of travel was not to be thought of except in case of dire necessity.

A good Samaritan appeared on the scene in the person of an exceedingly dirty and rowdyish looking young fellow, with the guise of a canaller.   He loudly invited every one to take the packet boat for Oil City.   Here was hope—doomed, alas ! to be crushed as soon as born. The packet boat was an oily flat boat, without shelter or seat, and the fare for the twelve miles by this precious conveyance was only three dollars and a half, or about thirty cents a mile! So that plan was rejected, and a brief council of war resulted in the decision to stay all night at Shaeffer's Farm, and start down on foot early in the morning. By dint of considerable finessing we secured a half interest each in a small bed packed with another bed in a dark closet dignified with the title of a " sleeping apartment." As four persons occupied the room, in which there was barely space enough for two persons to undress at one time, and as there was not a window or opening of any kind for ventilation, but little clothing was required to keep us warm, a fact of which the landlord was evidently aware. Our neighbors " across the way" were deep in oil, and kept up a continued conversation on the subject. About two o'clock in the morning I dropped asleep, lulled by a confused sound of " flowing well—five hundred thousand dollars—one half the oil—two years ago he wasn't worth a red cent—two thousand dollars a day—the biggest thing yet—third sandstone—made his everlasting fortune." My last mental reflection was that I wished I could say so of myself.

DOWN OIL CREEK TO OIL CITY.

He who essays the "middle passage" between
Shaeffer's Farm (the present terminus of the Oil
Creek Railroad) and Oil City, must prepare himself
for an experience for which life in the city affords
but a poor preparation. The second step from the
hotel at Shaeffer's plunges the pedestrian into a sea of
mud which extends with varying depth to Oil City,
more than twelve miles, with scarcely a friendly
rock on which to rest the sole of the foot. Mud
everywhere, illimitable, unfathomable. Let him
who thinks he can make the passage by turning up
his trowsers over his ancles and picking his way, at
once disabuse himself of the idea. If he does not, ten
steps from Person's Hotel at the Shaeffer will do it
for him.

Lest any intending visitor to the oil regions should
be discouraged by this picture and confine his wan-
derings to the limits of the railroads, I warn him
that if he would see anything at all of oildom he
must make the passage, unpleasant as it may be.
There is no alternative. To see the tips of the elephant's ears, or the end of his tail, is not to see the
animal, or form any idea of his bulk, and there is no
other way of doing it than to "wade in." And this
much may be added that whoever makes the trip,
with his eyes open, will never regret it. The sight
is one of which no description however graphic or
minute, can give a just idea.

The best way of making the passage, whether in the muddy season, or in the season of ice, is to travel on foot. It is the most independent, enabling the visitor to pursue his investigations with greater freedom, and is, moreover, in general the most expeditous way. A flat boat is an abomination, and a horse—especially such as they have on the Creek—is vanity and vexation of spirit. Strike out boldly on foot, and pull your legs up when they disappear from sight, remembering that if you descend deep enough you may strike oil. There is a choice of paths in going down or up the Creek, the difference between them being that each is muddier than the other, and that you are certain to select the muddiest.

The morning we set out from Shaffer's headed down creek, was intensely cold, with some little flying snow. The ground was frozen hard, with ridges and knobs, making the traveling even worse than it would be in soft mud. Not unfrequently what appeared to be solid ground would prove to be a mere thin crust, covering a deep mud hole, into which an unwary step would send the unlucky traveler knee deep, sorely to the wear and tear of Christian patience and forbearance.

After leaving Shaeffer's Farm the route lay through the Stephenson and Gregg farms.

With every rock and turn of the sinuous creek the derricks increased in number, and the wheeze and clank of the engines grew louder and more confused. Climbing around the bluffs, over a steep path, then striking the newly graded track of the unfinished Oil Creek Railroad, chiseled out of the face of the cliffs, and at last descending to the half frozen mud of the valley, we came out on the Foster Farm, crowded thickly with derricks and engines, groaning and creaking with the labor of pumping up the liquid treasures of the earth, more valuable than the golden

waters of the ancients. About sixty derricks were
massed together on this little tract of land, most of
them with their black, greasy vats, sometimes ranged
in a row, capable of holding each from five to ten
thousand dollars' worth of oil, and several of them
full. New derricks were going up, and engines stood
around, waiting to be put up in the proper places
and set to work. By the roadside, is a row of vats,
at one end of the row being the time stained and
greasy derrick of the famous Sherman well, whose history has already been given. Near it is the Porter
well, which in May, 1864, commenced flowing about a
hundred and fifty barrels daily, but now pumps from
fifteen to twenty barrels. On this farm are the wells
of the Briggs Oil Co., and across the Creek are those
of the Gillette Oil Co., both under the management
of Mr. J. T. Briggs, who has already been mentioned
as managing the property of the Indian Rock Oil Co.
The latter Company has purchased a large interest
in twenty-five acres on the Foster Farm, on which
are a number of producing wells. Most of the wells
were yielding finely, giving a large revenue to the
proprietors. The Briggs Company has been enabled
to pay heavy dividends from its receipts, and the
wells of the Indian Rock Oil Co. are producing sufficient to warrant large dividends on the Company's
stock. The Gillette Company's wells, on the Espy
Farm across the Creek, are partly producing and
partly boring, the prospects being very good for the
proprietors. Close to these wells are the old Buckeye
and the Buckeye Belle wells. The Buckeye formerly flowed largely, and bears an extensive reputation
from the fact that it was the oil of this well which
Mr. Briggs shipped to Europe as a sample, being the
first American rock oil ever sent across the Atlantic.
Having lain idle for some time the well became
choked, but has been restored by an air pump and is

now doing well. The famous Noble well, which in 1863 flowed twenty-five hundred barrels daily, and still flows largely, is on the same side of the Creek as the Gillette Co. wells, a little above. There also is the Crocker well, struck in 1863, and flowing largely for a considerable time, but now pumping.

The system of transportation adopted for oil and fuel is flat boating on the creek. Four horses abreast are attached to a flatboat, which they haul up stream, the horses taking the middle of the creek. The bed of the stream is even and covered with loose flat shale rock, the water being up to a horse's belly. An Oil Creek flatboat generally holds from 80 to 100 barrels of oil, on which the freight up is from seventy cents to one dollar, freight on coal being in proportion. As the boats sometimes make two trips a day, the business is highly profitable, though anything but pleasant, especially to the horses. As we passed down the creek the weather was intensely cold, and the ice was floating down in large masses, but the unhappy horses had to wade up with their heavy loads, their bodies partially clad in icy coats of mail, and their tails mere bunches of icicles. If it is borne in mind that these horses had to be from three to four hours in this icy water, without relief or rest, and that even saddle horses have to wade the stream several times in making the journey, the short lives and wretched character of the live stock in that region will not be wondered at.

Passing one or two "runs" with derricks going up or wells going down, we strike the McElhinney Farm, on both sides of the creek, punched as full of holes as a strainer. Here is the famous Funk well, the first flowing well on the creek, that kept up its stream of wealth for fifteen months, and, close by, the Empire well, that gushed forth three thousand barrels daily, and flooded the land around with oil a foot deep.

The Funk well is now silent and its lips dry, but the old Empire, after two years of steady flow, followed by a pumping yield of about a hundred barrels daily, and then a stubborn refusal to give another drop, has been induced, by the gentle persuasion of a "blower," or air pump, to send up about fifty barrels a day.

Passing through the Boyd Farm, on the East side of the Creek, on which there were fewer wells than on the tracts on either side of it, we crossed the stream to the G. W. McClintock Farm, where the throng of derricks, the clustered houses, and the flag pole in front of a tavern, marked the presence of " Petroleum Centre." Here the wells are crowded as thickly as houses in the most populous part of a city, dwellings and engine houses being mixed up in such inextricable confusion that it is difficult to distinguish one from the other without entering, and not always then. A ravine enters the Creek at this point from the West, and near the mouth are several producing wells, among others the Wild Cat wells, on the Gillespie property.

Re-crossing the river to the East side we came out on the Hyde and Egbert Farm, one of the most noted parts of the Creek from the number of important wells on it, among them being the Coquette, which is now flowing a large fortune into the pockets of its owners. Like most of the original owners of property on Oil Creek, both Dr. Egbert and his partner were men of small means before they struck oil, though both are now very wealthy. Several of the great flowing wells of the Creek are on their farm, and as the land owners have an interest—generally one half—in the oil raised, a comfortable income is secured. Here is the Maple Shade, which for months ran a thousand barrels daily, now dropped to fifty ; the Jersey, which still flows three hundred and twenty, three Keystone wells, flowing; a number of valuable pumping wells, and lastly the Coquette, struck a short time since, commencing to flow 1500 barrels daily, then falling to 1000 barrels, and now running 600 barrels.

The Coquette, being one of the latest great " sensations" on the Creek, is the object of much curiosity, and many pilgrims come daily to gaze in wonder and ‾ envy on it. Plashing through a quagmire three parts thick oil and one part mud the row of huge tanks was reached, and the rush of the stream of oil could be heard distinctly overhead. Climbing the derrick a view could be obtained of the interior of the tank, into which a stream of oil was rushing at the rate of five dollars a minute. Owing to the volume of gas in the well the oil is driven out like spray with such violence that at first it blew entirely across the tank, and saturated the ground around. A covering of boards has now been placed at the mouth of the pipe, and against this the stream plays with a force resembling the stream of one of the steam fire engines striking the side of a house. The quality of the oil was at first not of the best kind, being "riley" and of a yellowish color instead of the dark green of pure oil. It has now greatly improved and sells at full prices. There are twelve tanks, holding from 1200 to 1600 barrels each, full of oil, worth in the aggregate over $150,000, in addition to what has been barreled and sold. All around are notices warning visitors against smoking, the air being full of highly inflammable gas. A share in the Coquette is considered a "moderate" fortune. In January of this year Dr. Egbert sold one-twelfth interest in the Coquette for $250,000. Four years ago he bought the entire forty acres on which these flowing wells are located for one thousand dollars, taking his last dollar to pay the sum.

Passing the Rhinoceros well, the Porcupine well, the Ram Cat well, and a whole menagerie of other wells, we came to the Story Farm, crowded with derricks and wells. Here are the locations of the Columbia and Dalzell Oil Companies, two noted Pittsburgh Companies, the former being one of the most successful in the oil regions, returning larger profits to its original stockholders than any other company. Its history is such a remarkable instance of profitable investment that it will be read with interest. The Columbia was organized in 1862, and purchased the Story Farm for $128,000 from a company of seven persons, of Pittsburgh, who in 1859 bought the farm for a few thousand dollars. The Columbia Company was organized with a capital of $200,000, divided into 10,000 shares of a par value of $20 each. During the year 1862 the stock varied in price from $2 to $10 per share. At this time the chief difficulty with the company was the receipt of 1200 barrels of oil per day and no market for it. But a foreign demand soon sprung up, and between 1862 and 1864 the Company divided $300,000. In April 1864, $70,000 was divided, in May following $100,000, and $100,000 in June. The dividends between July and December were $625,000, making a total of dividends since the formation of the company of $1,195,000, more than five times the amount of original capital. In June, 1864, the old shares were called in and new ones issued of $50 each, the holder of an original $20 share receiving five new ones, of $50 each. The person who paid one year and a half ago the par value of $20 ecah for one hundred shares, and has held his stock, has received $12,000 in dividends to December, and from the profits on the increase of capital made in June last, obtained an accession to his stock of four hundred shares, which shares are worth, with his original hundred shares, at present market prices, $42,500, making a clear profit of $52,500 in eighteen months. If he bought the original shares at their lowest price, $2 each, that profit was made on a capital of $200.

Next below the Story Farm, on the East side, is the Tarr Farm, on which is the famous Philips well, which flowed two thousand barrels daily for many months. The owner of the Tarr Farm in years past was a poor and uneducated man, who eked out a meagre livelihood by lumbering in addition to scratching the barren hill for a scanty crop. Poor as the surface crops may have been, the soil below has sent up products so rich that the lucky owner is now an exceedingly wealthy man, who lives in splendid retirement at a small town not far from Meadville. The farm is covered with wells, nearly all of which, if not the whole, are successful. Judging from a superficial examination, this appears to be one of the most successful territories on the Creek, as it certainly is one of the most muddy. Here we came across a team, the first we had seen below Shaeffer's. It was stuck in a mudhole, the fore wheels clear under and the hind wheels invisible to the hub. The teamster, who, judging from that portion of him above ground, was probably a six-footer, stood contemplating the situation with dismay. In passing we ventured the remark, " Mister, guess you are stuck." It was a daring remark to make under the circumstances, and nothing could be expected in response less than a volley of curses, deep and dire. That such a proceeding suggested itself to the mind of the teamster was evident by the look of his eye, but after revolving the whole matter he concluded he could not do justice to the subject, and with one look at the " stalled " team and another at us, he gave a heavy groan and responded, " Well, it looks like it !"

Through muddy flats and up steep hillsides ; past throngs of derricks ; by gushing flowing wells, and

creaking pumping wells; through the Blood Farm,
where the dilapidated, unpainted, moss covered and
time-stained house, in which the owner of the farm
lived in his days of poverty, is confronted by the
smart and showy boarding house erected for the use
of its employees by the New York company now own-
ing most of the wells, at last we reach the Rynd
Farm, and the mouth of Cherry Tree Run. Here
the wells again become very thick, and abundant evidences exist of a large number of them being productive.

The widow McClintock, or Steele Farm, lies below
the Rynd Farm. A large number of valuable wells
are on this property, yielding a splendid revenue to
the proprietor of the land, John W. Steele, who has
but just come of age. The Buchanan Farm comes
next in line, and at this point, near the village of
Rouseville, another of the villages born of the oil excitement, is the old Taylor well that once flowed
largely, but stopped after yielding sixteen thousand
barrels, when it was left to lie idle. It has recently
been purchased by the Grant Oil Company of Cleveland, and restored by pumping to about thirty barrels, yielding a large return to the lucky owners.
A few rods farther down, Cherry Run enters Oil
Creek. Resisting for the present the desire to ex-
plore this ravine, now the scene of so much excitement on account of the numerous successful wells recently struck, we passed on to McClintockville, on
the Hamilton McClintock Farm, a small village, partly perched on a high bluff, and partly on the low
ground on the other side of the Creek.

Here again was a throng of wells, most of them
highly successful, with several new wells, many of
which had "struck ile." The Baltimore and Venango Oil Company have seven wells. All of the wells.
are doing finely, and the full vats and large streams.

of some of them would have gladdened the hearts of the stockholders had they been there to see. Close by those wells was one which has been the scene of a curious streak of luck. The owner sank his hole to the third sand rock, but found nothing but water. He pumped diligently for days, but without finding a grease spot in his vats, and then abandoned his unproductive hole in disgust. At this point he was visited by the owner of a neighboring well, who had been reaping the benefit of his labors, the water drawn up from the unproductive well having relieved the adjoining well of the stream which had previously caused some trouble. The visitor offered to pay the unfortunate pumper thirty-eight dollars per week to keep his pumps going, and, rather than abandon his engine, the latter agreed, and set to work once more. Six months he kept at it, drawing pure water out of the hole, to the relief of his neighbor, and then he "struck ile," and has since been pumping steadily to his own delight and the chagrin of his neighbor, whose vein he has "tapped." A nice question of ethics is involved in this matter. If the pumper was hired to pump his neighbor's water, has he any right to pump his oil ?

In the middle of the river, below the McClintockville bridge, is an old well. Tradition says that at this point a spring of oil bubbled up, and the Indians were in the habit of coming there to skim the oil for medicinal purposes. Here, also, it is said the owner of the land gathered the oil by soaking a blanket in the stream and wringing out the oleaginous fluid in a bucket. A few years ago a well was sunk on the spot, but the brilliant expectations indulged in by the adventurers were never realized. It was not a paying investment.

The Clapp Farm has a number of wells, many of them successful, but none of great note. Just be-

low the southern line of this tract is Cornplanter
Run, coming in from the West.   Preparations have
been made for boring on that territory.   The Graff
Hasson Farm, next above Oil City, contained one
thousand acres, and was purchased in 1856 for $7000.
A short time since three hundred and twenty five
acres of it sold for $750,000.   It formerly belonged
to Cornplanter, the renowned chief of the Seneca
Indians.

Oil City at last. Oil City, with its one long, crooked and bottomless street.   Oil City, with its dirty
houses, greasy plank sidewalks, and fathomless mud.
Oil City, where horsemen ford the street in from four
to five feet of liquid filth, and where the inhabitants
wear knee boots as part of indoor equipment.   Oil
City, which will give the dirtiest place in the world
three feet advantage and then beat it in depth of
mud.   Oil City, where weary travelers think themselves blest if they can secure their claim to six feet
of floor for the night, and where the most favored individual accepts with grateful joy the offer of half a
bed and the twentieth interest in a bed-room.

Oil City is worthy of its name.   The air reeks
with oil.   The mud is oily.   The rocks hugged by
the narrow street, perspire oil.   The water shines
with the rainbow hues of oil.   Oil boats, loaded with
oil, throng the oily stream, and oily men with oily
hands fasten oily ropes around oily snubbing posts.
. Oily derricks stand among the houses, and the "town
pump," if there is such an institution, must pump
oil.   There are several productive wells in the city,
ranging from five to twenty barrels, and the citizens
are busy boring in their back yards, in waste lots, or
wherever a derrick can be erected.   The Linden
well, just above the Petroleum House, is remarkable
from the fact that it commenced to flow on the 10th
day of October, 1861, at the rate of twenty barrels

per day, and has daily yielded a supply that has not
varied five barrels during the whole period, and
appears to be as vigorous to-day as when first struck.
The growth of Oil City is something remarkable.
Until the commencement of oil mining on the Creek,
there was nothing at the junction of Oil Creek and
the Alleghany but a small store and a tavern or two,
frequented by the raftsmen who brought their rafts
into the eddy and rested awhile. In 1861 a settlement was established at the mouth of the Creek, and
several stores of various kinds put up. In the
Spring of 1862, Oil City was incorporated as a
borough. There are now seven dry goods and general variety stores, three milliners, two jeweller's shops,
three banks, four drug stores, two photographic gal-
leries, two hardware stores, five boot and shoe stores,
thirteen family groceries and provisions, two groceries and bakery, five shoemaking shops, eight doctors,
four law firms, eighteen oil dock proprietors, eight
refineries, three oil brokers, four builders, two paint-
ers, twelve hotels, six saloons, two commission ware-
houses, one oil pipe fitter, two tinshops, three
blacksmiths, two machine shops, one wheelwright,
three butchers, three lumber yards, and four
churches.

In traveling from Shaeffer's Farm to Oil City, and
not taking into account any of the "Runs," over one
thousand wells, old and new, are passed. Before
Spring arrives that number will be largely increased.

CHERRY RUN.

The fact that Oil Creek itself is not the only valuable oil producing locality, and that apparently
valueless territory may prove highly productive, has
been exemplified in the history of Cherry Run. A
year ago this property was almost entirely neglected,
very few derricks were erected on it, and the land
was held at comparatively low prices. Now there
seems to be no limit, to the sums asked and paid.
The principal cause of this excitement is the success
of the Reed and Criswell and other flowing wells
which "struck oil" on the Run during the Summer
and Fall of 1864. The Reed and Criswell well commenced flowing about a thousand barrels daily, but
soon dropped to two hundred and eighty barrels, at
which it has remained steady for several months.
The quality of the oil is very fine. Soon after
striking of the Reed well, some others commenced
flowing. The excitement became intense, the rush
was tremendous, and in a short time all the available
property on the Run was taken up at high figures.
Failing to secure the fee simple to the land, the next
object of the late comers was to secure leases, and in
order to obtain these the anxious oil-seekers were not
only ready to give half the oil, but to pay large
bonuses in addition. Soon the valley was planted
as thickly with derricks as it could possibly hold,
where a lease could be obtained, and even the steep
hillsides were bored by the pertinacious oil-seekers.

To the utter confusion of theorists who hold that oil can only be found on the flats, and to the triumph of those who hold the opposite opinion, several of the wells away up the hillside have proved successful, which furnishes another proof that the only reliable theory is that oil exists wherever it flows or can be pumped out of a well, in other words, where it is it can be found, and where it is not it will not pay for seeking it. Nothing additional is charged for this bit of valuable information.

The scene in going up Cherry Run is more full of excitement than anywhere on the Creek. The Run, at its mouth, is but a narrow gorge between steep hills, through which a muddy, rocky, and brawling stream wends its way. Pumping wells and flowing wells are planted thick along the narrow flat, and climb the hills on either side. The road is execrable at its commencement, the wagons sinking over the hub, and at a short distance from the Creek loses its identity in a number of deep ruts looping around in all directions where a teamster could force his team in hopes of finding a better track. The stream being small and rocky, water transportation is not available, and powerful teams, drawing wagons loaded with five barrels of oil each, go plunging and staggering in the mud, and among the rocks that form the bed of the stream. Every now and then a wagon breaks down, and then the perpetual chorus of shouts and oaths becomes intensified in spots, making, with the noise of escaping steam, the clank and jar of the engines and pumps, and the rushing of the steam, a noisy enter-tainment. Traveling on the Creek is bad enough, but the extreme of diabolical locomotion is not attained unless the tramp is taken up Cherry Run. The constant passage of teams not only cuts the roads into deep sloughs of mud, but makes the pedestrian keep bobbing around to escape being knocked over by them in their erratic courses.

About two miles up the Run is the Reed & Criswell, or Reed well as it is generally known, Criswell
having sold out his interest for a princely sum.   In
the vicinity of this well are several other flowing
wells, among them the Baker well, credited with one
hundred barrels daily; the Gruninger, an intermittent flowing well; the Yankee, flowing fifty barrels;
and, a short distance above, the Auburn, flowing
eighty barrels. The two acres, on which the Reed
well is located, was offered for sale, two years ago,
for $1500, but found no purchasers. It was lately
sold for $650,000. Next above the Reed well is the
Smith Farm, comprising fifty acres. Three or four
years ago the then owner offered it for sale at $250
over the incumbrances. It was afterwards sold for
$2,400, and resold a year since for $6,500. It is now
the property of the Cherry Run Oil Co., who have
done nothing of themselves to develop their property,
and have therefore been at no expense beyond the
original purchase, but who are receiving from leases
on the territory over four hundred barrels of oil daily,
in royalty. A new well was struck a short distance
above the Reed well in January, flowing 250 barrels
daily, without the sucker rods being pulled o
Beyond the Smith Farm is the McFate Farm,
which there is a number of wells either down or
going down.
    Here the region of wells may be said to terminate
for the present. Above this point derricks innumerable are planted along the valley hillsides, but
engines are scarce. Leases are taken and derricks
erected thickly to nearly ten miles from the mouth,
but nothing further has been done. The terms of
most of the leases made on this Run require that
derricks shall be erected within sixty days, and
engines be on the ground within ninety days from
the date of the lease, and that the work be then

prosecuted with all reasonable diligence. In order to secure the lease as far as possible, the derrick is in all cases erected within the required time, but considerable difficulty exists in getting the engines on the ground within the required time, owing to the great demand on the machinists for engines.

Nearly four miles up the Run is the Humboldt Refinery, a very extensive establishment. The shipping point of this refinery is on the Alleghany in Walnut Bend, and in order to facilitate transportation the proprietors have constructed a road over the mountains, at a considerable expense, and established a ferry across the river. Most of their crude oil is brought from the Creek in wagons, but a considerable quantity is pumped up in pipes from the Tarr Farm.

Plumer village lies a short distance beyond the refineries, and about four miles from the mouth of the Run, or seven miles from Oil City by the road. The struggle to get property on Cherry Run has been so eager, that land has been purchased or leased for boring purposes a considerable distance above. At the beginning of the year 1865, there were on Cherry Run, from Rouseville to Plumer village, one hundred and eighty-six derricks, and about fifty above Plumer.

CHERRY TREE RUN, WEIKEL RUN, CORNPLANTER
RUN, REED RUN, AND TWO MILE RUN.

———————•◦•———————

The Cherry Tree Run enters Oil Creek from the
North-west, on the Rynd farm. It is a ravine of considerable size, with abrupt and lofty banks near Oil
Creek and widening into a fine valley as it approaches
the quiet little village of Cherry Tree. Until very
recently no attention was paid to this valley as an
oil locality, but the great success of the investigations on Cherry Run led to a more careful examination of all the ravines in the neighborhood of Oil
Creek. There are now a number of experimental
wells going down, nearly as high up the Run as Cherry Tree Mill. None of the wells have as yet reached a sufficient depth to fully test their value.

Weikel Run branches from Cherry Tree Run a
short distance above its mouth. It is a narrow ravine, with steep banks, covered with timber. It has
lately become favorite territory from the belief very
largely entertained by experienced managers of oil
wells that large deposits of oil exist in it. The indications certainly favor this idea, the configuration
of the ravine giving it the appearance of a slightly
diminished copy of Cherry Run. Less than half a
mile up is the celebrated "great gas well," which
made such a noise in 1864 by its unprecedented explosions of gas. The gas vein was struck in May,
when it blew with such violence as effectually to put
a stop to all further attempts at working. The vol-

ume of gas was tremendous, and its violence so great that anything thrown on the hole was instantly jerked into the air. Its roaring could be distinctly heard for a hundred rods. For nearly six months it continued to blow gas without cessation or apparent diminution, until at last the hole was plugged after considerable trouble. As soon as it was plugged the gas forced a stream of water into a well sunk some distance up the Run.

In December the gaseous mainfestations ceased, and the proprietors are preparing to sink the well with the expectation of finding a large deposit of oil. Whether their expectations will be realized remains to be seen, but on the principle that there can be no smoke without a fire there is evidently considerable oil in the neighborhood, if not immediately under the well. Acting on this belief the land around has been taken up by companies who propose testing it thoroughly. A small patch of land near the well, containing a small strip of borable territory and the rest " set up edgeways" has been purchased for $9000 by a Philadelphia company, and immediately above the gas well is the property of the " Weikel Run and McElhinney Oil Co." If appearances and " indications" are good for any thing, the lands of this company are admirably situated. Its proximity to the gas well gives assurance of finding oil, with a strong probability that the oil reservoir from which the gas escaped is a little removed from the well, and therefore on the company's lands. It is generally found when gas escapes in large volume from a well, without being accompanied with oil, that the gas has forced its way through cracks and fissures from the reservoir of oil at a little distance rather than that the oil is in the cavity immediately under the hole. The company have secured some valuable property on the McElhinney Farm, including two producing wells, in conjunction with their Weikel Run lands.

A short distance beyond the gas well is another well going down, having reached 660 feet. The show of oil was very encouraging, and the borers were confident of finding oil in considerable quantity as soon as the third sand rock was reached. Other wells in various stages of progress are scattered along both sides of the stream, considerable activity being displayed in developing the property. About two miles up a saw mill spans the stream, and just before it the ravine forks, one branch heading due West. Above the fork both branches of the ravine remain as yet undeveloped territory, but have good surface indications. A good road now exists to the Creek, and the facilities for shipment can easily be increased whenever required. Besides the heavy supply of timber on the land, the hill between the two forks of the stream is said by the neighboring farmers to contain a valuable vein of coal.

Taking the road from Cherry Tree to the Alleghany River, Weikel Run was left behind, and we crossed Cornplanter Run, which enters Oil Creek on the Clapp Farm. Preparations are making along its banks and bed to bore in the spring, but as yet no developments of consequence have been made. A little beyond this point we left the road and came out on Reed Run, a branch of Two Mile Run, which comes into the Alleghany not far from Franklin. Here were abundant evidences of a determined search for oil being in progress. Fifty-six acres in fee simple of the property through which Reed Run flows, have become the property of the Cleveland Cherry Valley Oil Company, who have an engine on the ground, and full preparations made for sinking a well as soon as the frost leaves the ground. Next below, the Magnolia Co. of New York have a well down ninety feet with a good show of oil. A tract of forty acres adjoining the Magnolia Company's tract has been purchased by E. L. Dodd and D. S. Keyes, of Cleveland, for the purpose of boring for oil as soon as the season will permit. Reed Run, especially in that part occupied by the three companies mentioned, is a very attractive territory to those who have a good eye for desirable oil locations. The bottom land of the Run, and the second bottom a few yards higher, afford as good prospect for successful wells as can be found on any of the Creeks or Runs in the vicinity of Oil Creek. The Cherry Valley Company have determined to thoroughly test the value of their property in that neighborhood, and from the known prudence and good judgment of the leading proprietors, who are from the ranks of the best and most sagacious business men, there is but little doubt that the affairs of the company will be so managed as to achieve success—if success is to be achieved at all—with the least possible amount of risk to the stockholders. That there is oil in the Run can scarcely be doubted, as it is but a very short distance from some of the most productive localities of Oil Creek, and but a short walk from several noted flowing wells. Should oil be found in any quantity on the Reed Run, the property there will bear a high value, there being ample space for boring a number of wells. The Cherry Valley Company have also two wells on the Alleghany which will be described in their proper place.

The forty acre tract of Dodd & Keyes, below the Magnolia Company's well, has a large amount of boring territory, very desirably situated, and is especially valuable for having a fine vein of coal in the immediate neighborhood of the oil flat. Lower down and near the junction of the Reed with Two Mile Run, is a well, down 350 feet, with good indications of oil.

Two Mile Run has not yet any producing wel
but preparations have been made for boring
thoroughly in the spring.

THE ALLEGHANY RIVER, WITH HICKORY, TIO-
NESTA, HEMLOCK, PITHOLE, AND OTHER
TRIBUTARY CREEKS.

During a portion of the summer months, before
the oil excitement extended far up the river, a small
steamer occasionally ran up from Franklin to President, on the Alleghany River, but those desirous of
going higher up, had to seek some other mode of
conveyance. In the Fall and Winter months, no
boats run above Oil City. A road follows the course
of the river, with ferries at the points where the
jutting of the precipitous bluffs out into the stream
stops the way. To those unaccustomed to the region
of rapidly flowing rivers, these ferries are interesting
novelties. Two strong and lofty poles are firmly fixed
in the banks and across their tops is stretched a
stout wire or iron rod, the ends fastened to the rocks
behind. A "traveler" or pulley wheel is placed on
that part of the wire which crosses the stream, and
from this "traveler" a line passes to the ferry-boat,
which is a flat, clumsy affair, on which passengers,
horses and vehicles are jumbled together indiscrim-
inately. When a load is on board, the boat is pushed
out into the stream, and the force of the current carries
her over without paddling, or care of any kind, the
"traveler," in its passage across the wire, emitting an
eerie sound that echoes strangely among the wild
hills at evening, and proclaims to those who have
ears to hear, the need of a well of lubricating oil in
the neighborhood.

Those wishing to strike the river above Walnut Bend, can shorten the distance considerably by taking the road from Oil City to Plumer, about seven miles, and thence by way of Neillsburg to Tidionte, or striking across to President, reach the same point along the river road. The latter route is the most interesting , and generally is a better road for traveling.

A daily stage runs from Oil City to Plumer, but but beyond that there is no public conveyance. Unless capable of performing a long march on foot, the best course is to get a horse in Oil City, and set out early in the morning. It is between forty and fifty miles to Tidionte, and there being much to see, and Oil City horses not noted for speed or bottom, the hours wear away rapidly.

Riding up the steep hill side from Oil City, a fine view is obtained cf the lower part of Oil Creek with the crowd of wells on the broad flat through which the stream runs towards the close of its course. A good birds-eye view could be taken from this point, although not representing the busiest portion of the oil region, and some enterprising photographist may find this hint profitable.

About a mile before reaching Plumer, the road crosses Cherry Run, and the multitude of derricks in the valley and along the hillsides, testify to the favor in which the Run is held by oil adventurers. The Humboldt Refinery lies in the Run, to the left of the road, and all the details of its extensive area lie open for inspection as on a builder's plan. The road at this point is horribly cut up by the heavily loaded wagons conveying oil to and from the refinery, and is very nearly as bad as the principal street of Oil City. Climbing another low hill the village of Plumer is reached, and at the center of the village the road to President branches off to the right.

Very little can be said in favor of the road from Plumer to the river. The first part is bad enough, but on reaching Pithole Creek it becomes worse. The Creek well deserves its name, as it winds its way through a gorge, dark, deep and forbidding. The road winds along the face of the precipitous sides, the last part of the descent being very steep and exceedingly miry. Long before reaching the bottom the rushing sound of the waters can be heard. A mill spans the stream, the road being carried by a narrow bridge across the mill dam, and climbing the other side amid crags and boulders at so steep a grade that a firm seat and steady hand are necessary in making the ascent or descent.

Pithole Creek obtains its name from some holes or small caverns in its sides from which a mephitic gas arises. A dog held close to one hole expired in a few minutes, and a goose, put into the hole, died in three minutes and soon became corrupt. A stone thrown into one of these holes can be heard rattling from ledge to ledge in its descent, until the sound dies out, rather than stops. The existence of these gas exhaling caverns led several persons to sink wells in hope of finding oil, but, although some success was met with at the mouth of the Creek, no very encouraging results were obtained higher up until, about the middle of January, 1865, a well on the Holmden tract, about seven miles from the mouth of the Creek, and about five miles north-east of Plumer, in an undeveloped territory, struck oil and flowed at the rate of 250 barrels per day. This created an immediate excitement, and Pithole Creek was swarmed with speculators eager to buy or lease every rod of land in the vicinity. The success of this well has demonstrated the fact that large supplies of oil can be obtained above the line of Oil Creek, and has increased the expectations of up river oil seekers.

3

A peculiar circumstance connected with this Pithole well is the fact that it struck oil in what is known as the fourth sand rock, being the only well in the oil regions that has reached that stratum. The first sand rock was reached at 115 feet, the second at 340 feet, the third at 480 feet, and the fourth at 600 feet. At 608 feet oil was struck. The well was drilled in November, but was not tubed until the middle of the following January. As soon as tubed the pump was set to work, and after an hour's pumping the oil began to flow, with the sucker rods in the chamber, at the rate of 250 barrels per day, at which rate it has steadily continued. The well is the property of the United States Petroleum Company.

After leaving Pithole Creek there is a good road on either side of the river, all the way to Tidioute, the hills falling back, or being less abrupt as the ascent of the stream is made, and better farming land appearing on the bottoms and in the rifts of the hills.

Tidioute, in Warren county, is the highest point on the Alleghany where there are producing wells. Above that point there are several of what are known as "farmers' wells," sunk in 1860 and 1861 by hand to a shallow depth, and abandoned when the depression in oil affairs occurred. Some attention has again been attracted to these wells, and preparations have been made for sinking them deeper, and also for testing the oil producing qualities of .Big Broken Straw Creek, which enters the Alleghany above Tidioute.

The Economite wells, owned by a religious sect known as the "Economites," are nearly opposite Tidioute. Five producing wells yield an aggregate of about sixty barrels daily, of heavy oil, the depth of none of the wells being over 120 feet. The Economite wells are in the side of the steep bluff a little way up from the river. The "Bretheren" are putting down several new wells in similar locations. On the

flat across the river, some other parties have put
down a well 600 feet without obtaining oil.

About half a mile below is a well down nearly a
thousand feet, belonging to the Tidioute and Alleghany Co.   Only one sand rock has been passed
through, at a depth of 150 feet.   A good show of
oil and gas has been obtained, but the deposit of oil
has not yet been reached.   On the other side of the
river are some shallow wells, of the same depth as
the Economite wells, producing five to six barrels
daily of lubricating oil.

Entering Venango county, the first point where
active operations have been commenced is at the
mouth of West Hickory Creek, on the upper part of
what is known as the Hickory Town Flats.   The
land has been purchased by the Hickory Farm Oil
Co. of New York, their property taking in about
three hundred acres on both sides of the Creek, commencing at the river and running back in the form
of an oblong parallelogram.   The hills which close
in on the stream through the greater part of its
length, recede soon after entering the lands of the
Company and leave a wide flat, on which there is
abundant space for a large number of wells, should
there be found sufficient inducement, of which there
is but little doubt.   A short distance up the Creek
a well was put down at a moderate depth about three
years since by the owner of the land, and worked by
a rude contrivance, part hand and part water.
From this well, with such rude appliances, oil of
superior quality is produced in considerable quantity
whenever worked, proving the existence of a large
and easily accessible vein in the neighborhood.   Another well, just below, was also put down over three
years since, and a good show of oil obtained, but
before completion it was abandoned, owing to the
depressed state of the oil market and the troubled

condition of the country, at the time when nearly all the wells were deserted from similar causes.

Directly on the Company's property, and close to the mouth of the Creek, three Scotchmen, named McKinley, sank a well in 1861. They reached a depth of 233 feet and found a fine supply of oil, promising to yield them a rich return for their investment and labor. Just as they were making preparations for tubing it the war commenced, and the owners of the well became so much alarmed at the condition of affairs that they abandoned their undertaking just as it was on the eve of resulting in such a splendid success, and their lease became forfeit.

The Hickory Farm Oil Co. have entered with vigor on the work of developing their property. Three engines have been brought on the ground and derricks erected for sinking two new wells, besides drilling deeper at the McKinley well. A competent superintendent and force of men are on the ground and working busily. A great advantage possessed by the Company is the abundant supply of good fuel, about two hundred cords of hard wood having already been got out for the use of the engines. All the wells are located near the river, so that the oil can be shipped without expense on steamers that can lie alongside, adding considerably to the value of the product. The strike, at Pithole Creek, of the Holmden well, which is but a few miles distant from West Hickory, has greatly advanced the value of property, and speculators are eagerly seeking for a chance to invest. The Hickory Farm could be sold for one million dollars to divide among companies that are seeking territory, and numerous applications have been made for leases, paying a royalty of one-half the oil.

On the other side of the river, just opposite the

McKinley well, is a well down 340 feet, tubed, and ready to pump, with a fine show of oil. The intensely cold weather has prevented further operations. On East Hickory four wells are going down. About a hundred rods below the lower well of the Hickory Farm Co. a Pittsburgh company are putting down a well. All of the adjoining lands have been taken up at high prices, and in the spring there will be a large number of wells sunk in the neighborhood of the Hickory.

About a mile below West Hickory Creek, on the same flat, is the property of the Pittsburgh and Alleghany Valley Oil Co., covering about 229 acres, held in fee simple. The purchasers of this tract selected it with great shrewdness, by far the greater portion being of the character generally known as rich oil flats. The frontage on the river is about seven eighths of a mile, and through the property a stream finds its way to the river, the ravine through which it comes affording good boring territory back the whole depth of the farm. In 1861 a well was sunk on the property to the depth of 220 feet, with a good show of oil, but was abandoned on account of the low price of the product. As yet no work has been done on the property by its present proprietors beyond the necessary preliminaries for commencing operations, but the engines will soon be on the ground, and the work of properly developing the tract commenced. The extent of the borable territory will enable the company to sink a large number of wells without interfering with each other's supply, a consideration the value of which will be speedily recognized by those who have seen the difficulties and losses experienced on Oil Creek in consequence of the "interference" of wells in too great proximity. The stock of the company is $200,000, of which $20,000 has been appropriated for working capital, and the remainder invested in the real estate. The managers of the company embrace some of the strongest names in Pittsburgh, the President, Dr. Curtis G. Hussey, and the Secretary and Treasurer, Thomas M. Howe, being names of high repute in the mining, financial, and commercial world.

About half a mile below the Pittsburgh and Alleghany Company's tract is the Sowers Farm, on which is a well that was struck in 1861 and flowed largely, but which, like nearly all the wells in the country, was abandoned in consequence of the low price of oil. It has now been put in operation and rendered productive.

Tionesta Creek comes in from the East a short distance below the Sowers Farm, which lies on the West bank. A few scattered houses and a tavern fronting the ferry landing, form the village of Tionesta. Around the mouth of the Creek and along the banks of the river there are abundant evidences of oil speculation, past and present. Shallow wells, hastily abandoned in 1861, rear their time-stained derricks on every side, whilst workmen, busy getting out timber for new derricks, and eager, keen-eyed men, with traveling pouches strapped to their sides, out "prospecting" for desirable sites, show the revival of interest in oil matters. A number of islands stud the surface of the river from Tionesta, past Lower Tionesta Creek, down to within about a mile of President, and on many of these islands old derricks and new derricks rear their heads among the unshapely trees.

The village of President, with its large, new, smart hotel, and its respectable gathering of houses, marks the junction of Hemlock Creek and Porcupine Run with the Alleghany; above the village is the well of the Farrar Oil Co. The whole territory surrounding the village, and extending up Hemlock

Creek and Porcupine Run in one direction, and down the river a considerable distance on the other, covering in all 8,400 acres, is the property of the President Petroleum Co., probably the largest corporation yet in the oil field, having a capital of five million dollars. Three wells are on the property, one near the McCrea Run, having reached a depth of 400 feet, with good show of oil, and two others on the river front, just below, going down on a lease taken by the Heydrick brothers.

Like Pithole Creek, Hemlock Creek has some strong manifestations of gas, or mephitic vapor. A story is told of three young men going along the valley in winter and finding the snow melted around a hole in the ground. One of them, a notoriously profane fellow, swore it was an opening into hell, and that he intended warming his feet at the fire. His companions endeavored to dissuade him, but he sat on the ground and stuck his feet in the hole, swearing with horrible oaths that he would warm his feet there if he had to go straight to hell in order to do it, and thanking the devil for finding him such an opportune supply of fuel. In a few minutes he stopped talking, and when his companions dragged him away he was totally insensible from the effects of the gas. His recovery was very difficult.

Just below the President Co.'s tract, at the foot of a lofty bluff is the celebrated Heydrick Well, sunk three years ago by the Heydrick brothers, young men who lived on the land, and who early adventured in grease. The well flowed for a considerable time from four to five hundred barrels daily, and then pumped one hundred barrels. When oil fell to a mere nominal value, and an empty barrel was worth its contents in oil six times over, the well was allowed to remain idle It has now been started up again and is making from twenty-five to thirty barrels

daily. The farm was owned by two Heydrick brothers and a brother-in-law, who like most of the farmers in the oil regions, had enough to do to make both
ends meet. The well sunk on their lands has been
leased by the Farmers & Mechaincs Co., who pay the
Heydricks half the oil, a tribute which puts a comfortable sum daily in their pockets. Four years ago
a big flowing well was gushing out oil next to the
Heydrick property, under the management of a
Michigan Company. When the Heydrick well was
struck, the Michigan well stopped, and no attempts
have, so far as we could learn, been made to recover
the vein. The Heydrick well has flowed and pumped,
to the present time, over thirty thousand barrels.

On the west bank of the river, directly opposite
the Heydrick well, is the Henry Farm, the property
of Hussey & McBride, on which there are several
productive wells. One has been flowing and pumping with large returns for two years, and now yields
forty barrels a day. Another "struck oil" at the
depth of 400 feet, and is yielding a hundred barrels
a day. Still another recently struck oil and is giving
large returns. But the principal well was sunk in
1861, and at the depth of 242 feet, obtained a flow
of oil that bewildered the proprietors. The greasy
fluid gushed up at the rate of three hundred barrels
a day, and continued to flow for three months. What
to do with the oil was a puzzle. Barrels could not
be got to ship it to market, nor vats to hold it on
the ground. Oil was down in the market, bringing
but ten to fifteen cents a barrel. There were no refineries in the neighborhood, and like the man who
won the elephant in the raffle, the proprietors of the
flowing well were "put to their trumps" to know
what to do with their prize. A little ravine close by
was dammed up and the oil turned into it until about
an acre of pure oil covered the ground two or three

feet deep. At last, in despair, the tube was stopped
with a pine plug, but the grease oozed up and escaped. At last the tube was bound over with the never
failing seed bag, and then the oil burst through the
earth and escaped into the river. Since then the
well has remained plugged, but it is now to be deep-
ened and re-tubed, when it is expected the old vein
will be again reached.

Next above the Henry Farm, and lying across the
mouth of Culbertson Run, is the property of the
Beekman Oil Co, on which there are three wells,
having a good yield of oil. Beyond this, on the
McCrea Flats, is the Kelley well, now owned by the
Cleveland and Buffalo Petroleum Co. This well
pumps from twenty to twenty-five barrels of oil daily,
of a superior quality, and has enabled the managers
to declare two monthly dividends of one per cent.
each. Considering that the Company has been but
a few months in existence, this immediate large return on the investment speaks highly of the value
of the property. The well, together with two and
a half acres of land adjoining, affording room for
five other wells, is held by perpetual lease. The entire flat on which the Kelley well is situated, extending from the lower boundary of the Henry Farm to
a point a considerable distance up the river, has
proved to be highly productive oil territory, and the
fact that the wells in the neighborhood all struck oil
at a depth less than three hundred feet, renders the
sinking of new wells a matter of much less expense
than in the majority of oil localities. The proximity
of the wells to the river, increases the value of the
oil by decreasing the cost of transportation.

Besides the leasehold property, the Cleveland and
Buffalo Co. have a tract in fee simple of eighty acres
on Culbertson Run, a short distance above the property of the Beekman Oil Co., the tract taking in the

well wooded hill-side, on which is fuel sufficient for
the engines for a considerable time, and a fine river
bottom on which there is room for several wells.
Part of the tract is bounded by the stream, and part
takes in both sides. This part of the Company's
property promises to be the most valuable portion
when properly developed, as it soon will be, the surface indications being unusually rich, whilst the proximity of so many highly productive wells gives
ground for strong hopes of a "big strike." The
Company is mostly composed of Cleveland business
men, numbering among them members of the leading
firms in the produce and shipping trade, who have
gone into the enterprise as a permanent investment.

Near the property of the Cleveland and Buffalo
Company, on Culbertson Run, is that of the McCrea
Petroleum Company of Pittsburgh, a company organized with a capital of $80,000, of which $20,000 has
been set aside as working capital, and is now in the
treasury. The tract leased by the McCrea Co. is
partly on the opposite side of the Run to that of the
Cleveland and Buffalo Co., and also overlaps the
stream for a short distance. A well was sunk on
the property in 1861, reaching 280 feet with a fine
show of oil, when the break-down in the price of oil
occurred, and the well was abandoned. Two engines have been purchased and are on the ground, to
sink new wells, besides drilling the old one deeper.
The lease was effected on highly favorable terms,
paying a royalty of only one-quarter the oil, and applications have already been made for sub-leases on
condition of half the oil, giving one quarter clear to
the company, in addition to three-fourths clear from
their own wells. The character of the enterprise
may be inferred from the fact that a very small portion of its stock has been put on the market, the
owners considering they have too good a thing to

part with much of their interest. The directors are some of the best business men of Pittsburgh.

On the McCrea Farm, lying between the property of the Beekman Co. and the lands of the McCrea Co. and Cleveland and Buffalo Co., is the territory of the Eagle Co. of Philadelphia, which "struck oil" during the last week of January, in one of their wells, which is now running over fifty barrels daily. This strike has made the property of other companies in the neighborhood increase greatly in value.

Past Pithole Creek, with the numerous wells clustered around its mouth, some of them producing a fair yield, down to Walnut Bend and Walnut Island. All along, the numerous spires of smoke from engine houses, the creak and wheeze of engines, and the steady plash in the black and greasy vats, told the story of remuneration for faith and labor. On Walnut Island a hundred barrel well was struck early in January this year, and gives signs of increasing.

From this point down the bluffs increase in height and steepness, and the flats are generally of less extent. Derricks line the narrow path at the foot of the bluffs, sometimes climb part of the way up the sides, and are planted thickly wherever there is a moderately wide shelf, or where a stream makes an opening in the hills. At Horse Creek, on the East side of the river, the Ross Oil Co.'s well is pumping about twenty barrels per day, and some other wells on that side are doing more or less, among them being the Wheeler well, doing thirty barrels. On the West side of the river, after getting a short distance below Walnut Bend, is the Hulings well, pumping twenty barrels. The Phillips well, unproductive, having stopped its yield some time since ; the managers are putting down a new well. The Revenue well, opened three years ago, abandoned, and now running under new management about twenty-five barrels a day.   The gas from this well is used to save a part of the fuel in the engine fires.   About opposite Horse Creek is the Brady Bend well, formerly flowing, now pumping ; College and Kincaid and Porter wells, four years old, and now revived and pumping each eight barrels per day.   Kincaid new well, that pumped sixty barrels per day for the first three days of its working in November, and now doing finely. Harrington well, pumping twenty-five barrels.

Farther down stream is the well of Long & Gay, pumping at a depth of 530 feet, with a fair yield. Beyond are the wells of Purchase & Co., two wells sunk in 1861 and abandoned, now cleaned out and pumped with one engine.   One well commenced pumping in September, 1864, at the depth of 348 feet, yielding fifteen barrels, with considerable gas. The other was pumped two months later, and gives twenty-two barrels, from a depth of 517 feet.   Five minutes walk distant is a well down 530 feet with a fair show of oil.   The derrick bears the legend of " Oil or China," and the borer swore he would either raise the oil or send his drill " up " through some Chinaman's cellar floor.   Still nearer Oil City is the Alcorn Farm, on which are the two wells leased by the Cleveland and Cherry Valley Co.   One of these wells is down 600 feet, and has commenced pumping oil, with every prospect of yielding in the neighborhood of one hundred barrels daily as soon as properly worked.   The other well is down 200 feet, with a good show of oil.   These wells will undoubtedly prove a valuable adjunct to the other property of the company, and enable them to pay good dividends on the stock.

From this point to Oil City, about one mile, there are a number of old wells and new wells, the latter just commenced to go down, and some of the former recently cleaned out and prepared to be sunk deeper.

From Oil City to Franklin is seven miles. During the season of navigation a small steamboat runs between the two points, but when ice runs strongly the steamer finds the pressure too much, and does not run in opposition. The Franklin Branch of the Atlantic & Great Western Railroad is completed to Oil City, and the cars will soon be running. The Jamestown and Franklin Railroad charter gives the right to construct a line to Oil City, but as yet the right to build is all there is of it. One line of conveyance is in full operation during all seasons,—that of " Foot and Walker," and it was by this line I made the journey.

The scenery along this part of the Alleghany river differs but little from that above Oil City, which has already been described. Where the bluffs approach the river they tower up to a considerable height, rising abruptly from the water, and having their craggy sides partly covered with timber. Where the bluffs recede, there is, between them and the river, a strip of tillable land, sometimes a quarter of a mile wide, and then narrowing to a mere ribbon, which is at length terminated by the steep bluffs. These intervals of low land generally consist of two levels, one but little above high water mark, and the other a plateau from ten to thirty feet above.

All along the river bank, on both sides, are oil wells, some of them yielding successfully, and others not yet sunk to the oil basin. Most of the wells are sunk on the strip of low land immediately adjoining the river, but a few are on the plateau, and several along the base of the steep bluffs or on ledges a few feet up the face of the bluff. Several of the wells gave good evidence of a fair yield, the stream of oil being of paying size and good color. Few have as yet properly developed their property, not having gone down to the third sandstone, which here lies deep, but

-contenting themselves with the yield from the less productive second sand-stone.

From Two Mile Run to Franklin there are several wells either producing or going down. A short distance below Two Mile Run, and about a mile above Franklin on the South bank of the river, is the property of the Milton Farm Oil Co., lying in one of those bits of low land, formed by the recession of the bluffs from the river, and through it Milton Run finds its way to the Alleghany. The territory purchased occupies twenty or thirty acres, extending from the river to the face of the bluff, to prevent any other claim getting behind to tap the oil vein, but not including any of the hill tops, consequently it is all borable property, affording room for a number of wells. But little of the oil property along this part of the Alleghany has the advantage of "creeks" or "runs," depending solely on the fact that it fronts the river. In this respect the Milton Farm is favorably situated, having the river front, as well as both banks of "run" to bore into. The river frontage extends for a quarter of a mile, affording ample space for sinking twenty wells. The Company, whose headquarters are in Cleveland, have an ample cash capital for sinking three or four wells, which they are preparing to do at once. Among those principally interested in the company are Anson Stager, General Superintendent of Western Union Telegraph, Geo. B. Hicks, inventor of the Hicks Telegraph Repeater, H. Garrettson, Amos Townsend, Lemuel Crawford, and other leading business men of Cleveland. Three wells are going down on the land immediately adjoining, and on the same reach of low land. On the opposite side of the river, and a little below is a well that has been for some time producing a fine yield of lubricating oil.

The fact that all the wells along this part of the

river are only down to the second sandstone, makes
it evident that they can not give as large a yield as
the great flowing wells on Oil Creek, that are down
to the third rock. But the corresponding fact that
the second rock wells all yield lubricating oil, com-
manding more than double the price of the Oil Creek
product, is a complete offset to the smaller yield.
By boring from thirteen hundred to two thousand
feet, it is believed that the third sand rock can be
reached, with a greater flow of oil than can now be
obtained on Oil Creek. The correctness or error of
this assumption will be tested before long, as, lower
down the river, some well owners think cf sinking
two thousand feet, if the third rock is not found at a
less depth. The fact of the superior facilities for
shipment possessed by wells on the river, especially
within the limits of regular steamboat navigation, is
too self-evident to need argument. It is enough to
add that the expense of getting the oil to the place
of shipment is always taken out of the price of the
oil at the wells, and that oil produced on the river
will always therefore bring a higher price than the
same quality produced at a less accessible place.

Passing several wells in operation, old wells being
cleaned out and prepared for re-working, and new
wells boring, Franklin was at length reached, the
quaint old capital of Venango county, with its old
fashioned houses, its muddy streets, and its miserable
tumble-down Court-house, in which land sales of from
one to three million dollars a day are recorded, and
documents of incalculable value are stored without a
vault to protect them from the accident of fire. The
average consumption of revenue stamps in the Re-
corder's office is estimated at about $500 per day,
making a very handsome revenue to the government.

Franklin is a very old settlement, being the site of
three forts, Fort Venango, established by the French,

a fort built by the British, and Fort Franklin, built
by the Americans in the war of independence. In
1795, the town of Franklin was laid out on the site
of the last named fort, and afterwards became the
capital of Venango county.

It now contains a population of about thirty-five
hundred, and is a growing place. A suspension bridge
spans the river at this point, the old bridge having
been burned down over eighteen months since, by
some blazing oil boats that took fire in the great con-
flagration at Oil City. Franklin is the present terminus of the branch of the Atlantic & Great Western
Railway, over which a very large business is done.
The Jamestown & Franklin Railroad, partially built,
will also connect Franklin with Cleveland and Buffalo,
by way of Ashtabula and the Lake Shore Railroad.

There are several wells in operation within the
borough limits, and the product is very satisfactory
to the owners. Below Franklin, the river is lined
with wells for several miles, many new ones going
down, and several old ones flowing and pumping.
Among the producing wells are the Keystone Well,
pumping about four barrels ; the Lee Well, about 500
feet deep, and flowing about fifty barrels ; the Dale &
Morrow Well, pumping about thirty barrels from a
depth of 450 feet; the Hoover-Island Well, on the
first island below Franklin, pumping and flowing
seventy-five barrels daily. A number of wells with
fair yield of oil are scattered along the river to a
distance of several miles. East Sandy, Big Sandy,
and Scrub Grass Creeks, flowing into the Alleghany,
are also occupied as oil territory, and operations have
been pushed vigorously on the Big Sandy.

FRENCH CREEK, SUGAR CREEK, AND OTHER OIL LOCALITIES.

From Franklin to Meadville, twenty-eight miles, the Franklin Branch of the Atlantic & Great Western Railway runs along the bank of French Creek, an important and pretty stream, considerably larger than Oil Creek, during the greater part of its length, and also deeper. For several miles up the Creek there are old wells and new wells, several of the latter producing oil; among the most noted being one on the Sutley Farm, a short distance above Franklin, and the well of the Tallman Co., near Utica Station. The oil produced on French Creek, being of a heavy lubricating quality, bears a higher value than that of Oil Creek. The land along nearly the whole length of the Creek has been purchased or leased for boring, and most of the abandoned wells of 1861 have been taken by new companies who have the capital and energy to properly test the property.

A favorite region at the present time is Sugar Creek which takes its rise in Cherry Tree township, Venango county, on the same tract out of which Cherry Tree Run flows. It passes through the borders of Plum and Oakland townships, in a S. W. direction, to Cooperstown in Jackson township, and then runs nearly due south through Sugar Creek township to French Creek, which it strikes about two miles above Franklin. It receives several branches, the largest of which is West Sugar Creek, which rises in Sugar Lake, just over the Crawford county border, and joins the main creek at Cooperstown. Through its whole course it passes through a fine farming country, the cultivated flats and hillsides and good roads affording in this respect a decided contrast to some portions of the oil regions.

The Creek is not large enough to afford water facilities for shipping oil, but a good road keeps along the flat valley to the junction with French Creek, four miles, where there is a station on the Franklin Branch of the Atlantic and Great Western Railway. In passing up the Creek, evidences of explorations for oil, and of the oil itself, become speedily visible. At the mouth, a well is located, and as the ascent of the stream is made, other wells in various stages of progress come into view. On the McCalmont farm, about two miles and a half above French Creek, is a well put down, about three years since, and worked by water power. In all, from three to four hundred barrels of heavy lubricating oil have been obtained from this well, such as now sells at $20 to $25 per barrel at the well. The hole was only sunk to the second sand rock, reaching a depth of 312 feet, none of the old wells on this creek having gone to a greater depth or penetrated beyond the second sand rock, the supply of heavy oil having induced the owners to stop at that point. A company from Rochester leased the well, put up an engine, and in a very short time struck a vein of pure lubricating oil.

From this point up there are several wells and derricks, but few producing anything of consequence until Cooperstown is reached. Immediately above the village, on the Booth and Hillier Farm, a well similar to that on the McCalmont Farm was put down 312 feet by water power, when it struck lubricating oil. The well has been purchased by two experienced parties from Oil Creek, who are confident of finding a good supply of oil in the third sand rock, toward which they are boring, having reached a depth of 600 feet. Two wells are going down on the Sweeney farm next above, one having got down 100 feet, and the other 300 feet, the latter striking oil at 80 feet.

Adjoining Cooperstown on the south, and partly bounded by the borough line, is the farm of the Sugar Creek Co. of Cleveland, containing one hundred acres, a large proportion of which is good boring territory. On an island included in the property, is a well put down three years ago by hand, to a depth of 297 feet, with a good show of oil. When the panic occurred the well was abandoned. It has now been leased and will soon be sunk deeper and tubed. Another well is going down on the property, and preparations made for a third well. A Company from Cleveland and Sandusky have purchased the Alexander farm, next below the Smith farm, and are preparing to sink wells.

From Cooperstown to Utica is but a short drive, and not a long walk. There the cars can be taken for Meadville, and thus the grand circuit of the Venango County oil regions be completed. The weary traveler will be glad to exchange the discomforts and hardships of his tour for the warm welcome and luxurious comforts that await him at the McHenry House.

In the foregoing pages a full description has been given of the oil regions of Venango county, Pennsylvania. There are, however, several other localities in the State where oil indications exist, and where the work of development has been commenced. In Warren county, the presumed oil territory lies along the banks of the Alleghany River and Big Broken Straw Creek. Upper Oil Creek in Crawford county, and Upper French Creek in Crawford and Erie counties, have been taken up for oil territory.

The Clarion River, through Clarion, Elk, Forest and Jefferson Counties, is the scene of a great oil excitement. South of Pittsburgh, in Fayette and Greene counties, numerous wells are sinking on Dunkard Creek. But in all these the operations are but commencing, and whether the developments will equal those of Venango county, remains to be proved.

WEST VIRGINIA AND SOUTHERN OHIO OIL REGIONS.

The oil region of West Virginia is an unmistakable and well defined geological formation, known as the "up-heaval" or "Great Oil Belt," extending from the Ohio River, opposite the little Muskingum and Duck Creek, about forty miles in a direction a little west of south, varying in width from three to ten, or perhaps fifteen miles. The rocks are peculiarly disturbed and broken; the hills, along the numerous streams and gorges, varying from one to three hundred feet high; and along the centre of the belt the rocks are nearly vertical, but dip at various angles as they recede on either side, forming what is called the East and West slopes. By some convulsion of nature, the rocks appear to have been "up-heaved," and separated, making deep ravines, gorges and gullies, many of which have become the permanent beds of streams, along the bottoms of which is found the "boring territory" as indicated by the color and character of the rocks, and the presence of oil, both on the surface and oozing from the fissures of the rocks.

The principal streams arising in, and running through these gorges or openings, and penetrating the Great Oil Belt, are the Little Kanawha, Hughes River, with its North and South Forks; Goose Creek, with its Laurel, Pigeon Roost, Myers, Oil Run, Ellis', Buffalo, and First and Second Big Run Forks; Mill-Site Run; Walker's Creek, with its Straight Walker

Fork, Silver Run and Bee Tree Run Forks; Stillwell
Creek, with several forks; Bull Creek, with its Horse
Neck and Isaac Forks; Cow Creek; Calf Creek; Rawl-
son Run Fork; French Creek; Standing Stone Creek;
Burning Spring Run, and other streams of less note.

This Oil Region is completely undeveloped, yet
the existence of Petroleum or British Oil, as it was
called by the settlers, has been known for more than
fifty years. Thousands of barrels of oil have been
taken from pits sunk in the sand on the banks of
Hughes River. In 1860–1 the high oil fever existing
in the Venango Valley of Pennsylvania, spread out
to this region, and several enterprising companies
and individuals commenced boring for oil, on the
Hughes River at Oil Springs, and on the Kanawha
at Burning Springs. At the former place a flowing
well was struck, which up to to-day, continues to flow
at the same rate as when first opened, from two to
six barrels of oil per day. At the Burning Springs,
the great Llewellyn well was struck, which flowed for
several months at the rate of from fourteen hundred
to two thousand barrels per day, and is still largely
productive. A great many wells were commenced
at different localities; some on Cow Creek, Stillwell,
Oil Creek, Walker's Creek, and the creeks near the
Burning Springs, all of which, that were not broken
up by the Rebels at the beginning of the Rebellion, produced oil in greater or less quantities.
The fifteen or sixteen wells on Oil Creek, at Petroleum, yielded, and still continue to yield, from two
to three hundred barrels of superior lubricating oil
per month, their depth being only from 80 to 160
feet.

On account of the Rebellion, all operations were
suspended from 1861, up to Sheridan's successes in
1864. Since then this region has been the theatre of
the most intense excitement. Experienced oil men

from Pennsylvania have secured large tracts of oil territory; numerous enterprising companies have been formed, securing hundreds of acres of the choicest boring territory, and to-day, where but recently all was so still and desolate, may be seen nearly three hundred derricks, and the cheerful puffing of as many engines.

It is a fact, worthy of consideration, that not a single well has been abandoned on account of the failure or non-appearance of oil, while in almost all other oil regions, a large number have been abandoned as " dry wells." It is believed that no greater or more productive wells have been opened in any region, than in West Virginia, and in proportion to the time spent and capital invested, no where has the enterprising oil seeker found a more sure and abundant return. The development in Pennsylvania began about ten years ago, a vast amount of capital has been expended, and nothing has occurred to retard the vigorous prosecution of explorations. The success there has been very great, almost fabulous. In West Va., the enterprise is but just begun, yet all experienced oil men, and the most skilful geologists, concur in the belief that the same time and similar enterprise will develop here an equal, if not a greater yield of oil. At Horse Neck there are some twenty wells, producing from ten to sixty barrels per day. D. H. Wallace, in company with the Philipses of Oil Creek, own about 5,000 acres of oil territory on and in the vicinity of Bull, Cow and Calf Creeks, and have some sixty wells in operation, or in process of boring, with suitable engines. There are several wells going down on Stillwell, one on Walker's Creek, by Mr. Murray, near the Smith farm, having a fine show of oil at the depth of 160 feet, and having at 30 feet gone through a stratum of copper ore, of superior quality, about 13 feet thick. Mr.

Candy has two wells in progress above the Petroleum Wells. The Great Belt Oil Company, of Cleveland, have one well down over three hundred feet on the Hall farm, and one 260 feet on the Sharpnack farm, near the "Oil Springs," with oil sufficient to warrant tubing. The wells on Oil Creek, known as the Petroleum Wells, are doing finely, considering the fact that fifteen of them are pumped by one engine. That Company is putting down several new wells, with every show of success.

The Baltimore Company, Mr. Cannon, President, are prosecuting their works with becoming energy near the Oil Springs, on Mill-side Run, just below the Sharpnack Farm of the Great Belt Oil Company, where they have one well down over five hundred feet with every prospect of success, and two more engines and derricks up and nearly ready for work. The wells in the vicinity of Burning Springs are all making a good yield, and next spring a large number of wells will be put down on the Rathbone tracts and on the Standing Stone, by several enterprising companies.

From the number of working companies formed, it is safe to predict that during the coming year rich developments will be made, and great wealth accumulated in the region of the Great Oil Belt of West Virginia.

The region of country across the Ohio River, in Ohio, in the line of the Great Belt, of West Virginia, is being rapidly taken up and developed, and bids fair to prove no insignificant rival to the better known localities of Oil Creek, Pa., and Kanawha Valley, West Va. Five paying wells have been struck on Duck Creek, Wolf Creek and Buffalo Creek, and further discoveries are constantly making as the drills, in different localities, go down.

The oil found in West Virginia and Southern Ohio

is of a superior quality. Most of the shallow wells produce a lubricating oil, very heavy and of great value, the illuminating oil having none of the offensive odor that is sometimes found; it is of excellent quality for refining. The lubricating oil commands readily $30 at the wells. Geologists and experienced oil men concur in the belief that the illuminating oils will be found by sinking the wells producing the lubricating oil to or through the third sand stone.

So far as has yet been discovered by boring, the description and stratas of rocks in the West Virginia region are the same as those in Venango county, Pennsylvania. The upheaval is composed of a reddish brown sand stone. The strata below, as found by boring in the valleys of the streams, are about as follows in both regions, varying somewhat as the break or the slopes of the Belt are approached: first sand stone from 30 to 200 feet; soft rock or shale, from 10 to 100 feet; second sand stone, from 50 to 150 feet; shale or soap stone, 10 to 30 feet; third sand stone, from 60 to 100 feet : shale or soap stone, from 30 to 50 feet; fourth sand stone, from 60 to 100 feet; lime stone.

THE OIL REGIONS OF LIVERPOOL, OHIO.

Attention has recently been attracted to the neighborhood of Liverpool, in Medina county, Ohio, as a place for the production of petroleum. Oil of a superior quality had been known to exist there for many years, and small quantities have been produced and used for various purposes for nearly half a century. In the winter of 1860-61 several wells were sunk for the purpose of obtaining oil, and some of them yielded a fair supply of a very superior quality. None of them had been properly tested when the discovery of the flowing wells on Oil Creek, and the consequent glut of the oil market, caused the abandonment of all wells producing in small quantity, and the suspension of all experiments in developing the value of oil territory. From that time until the present no systematized effort has been made to produce oil from the wells already sunk at Liverpool, or to make a more thorough test of the extent of the oil deposit.

To reach Liverpool, take the train from Cleveland to Grafton, 26 miles, when a ride of nine miles on a good road brings the visitor to his destination. The road from Grafton to Liverpool is mostly over level table land, the surface soil being clay, giving a good road in dry or very cold weather, but very heavy traveling after rain. A few slight depressions on the way break the general uniformity of level, and in most of these depressions there are greater or less indications of oil.

When near Liverpool the ground drops abruptly
to the flat valley through which the West Branch of
Rocky River flows in its serpentine course. At this
point and for about two miles below, the valley is
wider than above or below, forming a hollow lake-like
basin. The immediate banks of the stream are from
six to ten feet high above, this being a dead level
until the boundaries of the valley are reached, where
the banks are from thirty to sixty feet in length,
increasing in height and abruptness towards the
lower end of the basin. The village of Liverpool
stands at the upper end of the basin.

Below the village a short distance is what is
known as the "Oil Spring," being a salt well sunk,
in 1819, about a half a mile below the village, on the
west bank of the river. The well struck the sandstone at the depth of ninety feet, when oil came up
accompanied with explosions of gas that terrified the
borers. The well was bored still deeper, reaching in
all a depth of 160 feet. Salt water was found, but
the salt was in such small proportion as not to be
profitable, and the large admixture of thick oil dam-
aged the quality of the salt.

A hollow wooden tube was sunk in the well, and
through this oil came up, accompanied by frequent
rumblings and explosions of gas. A hand pump was
used, and sometimes two barrels of oil a day were
obtained. The spontaneous flow of oil enabled the
owner of the well to dip off about a barrel in a week.
A number of years since three barrels of the oil were
taken to Cleveland and offered for sale at the drug
stores as "rock oil," to be used as a cure for rheumatism, but the quantity was too great, and no one
would buy it. It was used in the country around
Liverpool as a "sovereign cure for the rheumatics,"
also as a remedy for hoarseness and sore throat, and
as a lubricator for machinery and cart wheels.

In the winter of 1860–61, a number of experiments were made along the valley for the purpose of developing its value as oil territory. At Marysville, a hole was sunk two hundred feet by G. V. Harper. The drill passed through two layers of soapstone and one of sandstone. A good show of oil was obtained, when the owner of the well conceived the idea that by blasting the rock at the bottom of the hole he could shatter it and form a cavity into which the oil could flow, and then be pumped up. He put down a gas pipe, through which he dropped powder and then exploded it. The rock was shattered, and so was the gas pipe, which remained firmly fixed in the hole, and is still there. The force of the explosion drove up the oil into the river in a number of places, but whether it accomplished the object of making a cavity and filling it with oil is a question that has not been solved.

Near the Harper well, on land belonging to Thomas and William Purdy, the late Colonel Heman Canfield, of Medina, bored a hole in the bed of the river, and struck oil at the depth of ninety feet. He put in a gas pipe and pumped the hole by hand, getting about a barrel per day. A flood soon after raised the stream, and the running ice bent the pipe and stuck it fast in the hole, so that it could not be worked. Col. Canfield abandoned the river well, and sunk a second well on the bank, reaching the sandstone at the depth of 120 feet. A small quantity of oil was found.

Four wells were bored by Col. Whittlesey, neither of them being driven into the sand stone. A large shaft was sunk at another point, to a depth of sixty feet, and a hole bored from its bottom to an equal depth below. Heavy oil was found, and about thirty barrels obtained, when the price fell and the well was abandoned.

Another well was sunk near the last well of Col. Whittlesey—or the Deming well, as it is now called—by Messrs. Carr and Tillotson. This hole was put down 110 feet to the sandstone, but not finding the oil in the shale rock, where it was then supposed the deposit existed, the work was stopped.

In the heart of the village of Liverpool a tinman named Born sank a well in his back yard to the depth of 132 feet, when oil was struck. A hand pump was put in, and subsequently a very small engine was used. A number of barrels were obtained in this way, when oil fell to such a low price that only five dollars a barrel could be got, including barrels and hauling to market. The pump was taken out and the engine sold. The oil now lies in considerable quantity on the surface of the water in the well, and Mr. Born dips it up by attaching a tin vessel to a pole. He sells it to the village neighbors and the farmers for lubricating purposes.

About two miles above Liverpool, a well was sunk by Howard & Ford to the depth of 130 feet. They abandoned it before reaching the sand rock, although good indications of oil were found.

An old man named Farnsworth, a machinist, having a shop in the side of the hill to the south-east of the village, bored a hole in the bed of a small stream, sinking it 212 feet. The drill passed through the upper crust of sand stone, and struck the shale rock below. Considerable gas came up, but no oil in any quantity.

In summing up the results of these experiments it appears that of the thirteen wells, in all, sunk in what may be called the Liverpool basin, and near it only six penetrated the sandstone, the others stopping at the sandstone, or before reaching it. Of the six that penetrated the sandstone, five found oil; the sixth passed through the first sand stone into shale, and there stopped, finding gas, but no oil. One of the five that found oil entered slightly a second layer of sand stone. The first sand stone was struck at a depth of 90 to 132 feet. The deepest well sunk was 212 feet. No effort has hitherto been made to penetrate to the lower beds of sand stone, in which it is believed by many that large deposits of oil exist.

That oil of a very superior quality exists in the Liverpool basin, there is ocular and olfactory evidence. It is forced up by the gas at the old oil spring; it floats to the depth of several inches on the water in the Deming well; and the people of the whole country-side use it commonly. It is brownish-black in color, and about the thickness of ordinary molasses. Samples have been shown to Cleveland oil dealers, who offer $50 per barrel for it as a lubricator. At this rate, a well producing three barrels daily would pay a handsome profit.

Vigorous preparations are making for the proper development of the oil resources of the basin. The pioneer in the work is the Liverpool Oil Company, of Cleveland, which has effected leases on favorable terms of between four and five hundred acres, having a frontage on the river and the tributary streams of about three miles. On their territory are three of the holes bored by Col. Whittlesey to the sand stone, and the large well sunk by him, now known as the Deming well. At this well their first operations have commenced. After the water in the hole had been pumped out several barrels of heavy lubricating oil was found, but the engine was too small to operate properly, and a new and more powerful one was ordered. In the mean time the engine has been set to work drilling a new hole a short distance from the present well. Another well is sinking near the "Oil Spring Lot," with fine show of oil.

The extent and character of the territory covered

by the leases of the Liverpool Oil Company afford abundant opportunity for thoroughly testing the resources of the basin, and the prospect so far as can be judged from the experiments already made, is very promising. In order to settle the question as to the existence of large deposits of oil in the lower sand stone rocks, some deep wells will be sunk. The terms on which the leases have been obtained are highly favorable, and the company have capital to prosecute the work with energy. A large number of the active and able business men of Cleveland are interested in the company and direct its affairs.

Besides the Liverpool Oil Company, over twenty different companies and individual adventurers have made purchases and leases for the purpose of thoroughly developing the oil resources of the locality. The old salt well has been purchased with other property, by a company, and its value as an oil well will soon be tested. Leases and purchases have also been effected in the neighboring township of Grafton, and there is every prospect that before many months the valley of Rocky River will be planted as thickly with derricks as some of the most productive ravines of Venango county.